A Primer on Tutoring Mathematics

Robert L. Jones

INFINITY
PUBLISHING

ISBN 978-0-7414-6305-0 [illegible]
ISBN 978-0-7414-6306-7 [illegible]
ISBN 978-0-7414-9290-6 [illegible]

Printed in the United S[illegible]

Published March 2012

[illegible] BLIS[illegible]
10[illegible] en Str[illegible]
West [illegible] en, P[illegible]
[illegible] BUY[illegible]
[illegible] (610) 9[illegible]
[illegible] 941-[illegible]
[illegible] bonBo[illegible]
[illegible] son[illegible]

Preface

Low-level math is based on simple concepts, and uses only six mathematical operations: addition, subtraction, multiplication, division, and multiplication and division by means of exponents. But math is presented in equations, which are like a foreign language to many students. Students can understand math; they just can't read it. Here are a few equations that we derive in this book. They are unintelligible to the general public.

$$D = D_0\left(1 + \frac{r}{n}\right)^{nt} \qquad \lim_{n\to\infty}\left(1 + \frac{1}{n}\right)^n = e \qquad x = \frac{-b \pm \sqrt{b^2 - 4ac}}{2a}$$

$$(a + b)^n = \sum_{k=0}^{n} {}^nC_k\, a^{n-k} b^k \qquad A = \begin{vmatrix} a_{11} a_{12} \\ a_{21} a_{22} \end{vmatrix} \qquad AB \neq BA$$

$$e^{i\theta} = \cos(\theta) + i\,\sin(\theta) \qquad P = S\,|\cos\varphi| \qquad \frac{1}{1-x} = 1 + x + x^2 + x^3 + \ldots = \sum_{n=0}^{\infty} x^n$$

$$\frac{d(\tan\theta)}{d\theta} = \frac{1}{\cos^2\theta} \qquad dy = d\left(\frac{u}{v}\right) = \frac{vdu - udv}{v^2} \qquad f(x) = \sum_{n=0}^{\infty} \frac{f^n(0)}{n!} x^n = f(0) + f'(0)\cdot x + \frac{f''(0)}{2!}\cdot x^2 + \ldots$$

This tutorial teaches students (and parents) to read and understand equations by reviewing the basics of mathematics - arithmetic, exponents, rates and interest, problem solving by graphs, basic algebra, matrix algebra, trigonometry, the Fibonacci sequence as well as the power, Maclaurin and Taylor series approximations, and first-level calculus. Each review is short, but written in a way that gradually introduces the symbols and conventions of mathematics, and teaches students to comprehend equations.

The book uses a " .. how did they do that?" approach that is easy to follow, and more importantly, to remember. It also provides clear insights into how math works, as well as numerous "trivia" points (see back cover of book) to pique students' interest. My goal is not to train new mathematicians - that is far beyond my abilities. Rather I hope this tutorial will "bring math down to earth", and make the use of mathematics easier for the average student.

The book is intended for students beginning high school mathematics, their parents, and especially parents planning to home school their children in math. It may also be helpful to lower-level math instructors since it gives "plain English" explanations (useful for students) of the basic operations of mathematics.

About the author

Dr. Jones holds a B.A. in Chemistry from Southern Illinois University (1958) and a Ph.D. in Physical Chemistry from the University of Michigan (1963). He was a Research Chemist at the U.S. Naval Research Laboratory for 34 years where he studied such topics as boiler corrosion, stress-corrosion cracking, molten salt hot corrosion in gas turbine engines, high temperature reactions, phase stability in zirconia-based thermal barrier coatings, and catalytic combustion. Dr. Jones has published more than 50 articles in scientific journals, and has been granted seven patents. He was the recipient of the NRL-Sigma Xi Award for Applied Science in 1986.

Dr. Jones is also the author of the book, *Common-Sense Thermodynamics: A Tutorial on Understanding Thermodynamics* (Nova Science Publishers, NY, 2002).

T[illegible] C[illegible]

Introduction

One would think that the tutoring of mathematics should be easy, or at least not difficult. Low-level math is based on simple concepts, and uses only six mathematical operations — the four arithmetic operations plus multiplication or division by means of exponents. And all of these operations can be performed quite easily on any scientific calculator. But there are two main problems: 1) mathematics is presented using symbols and conventions that are not familiar to beginning students, and 2) and it involves equations which many students can not "read." They do not know what the equation is saying.

The purpose of this book is to provide a broad review of basic mathematics for students and parents undertaking the study of mathematics. The book is small, and meant to be easy to read. Mathematics should not be a contest of how hard the student (and parent) can work. Rather it should be, if approached from the right viewpoint, more a matter of just curiosity and intellectual interest. Our book's essential goal is to give the student the *vocabulary* and some of the *sense* of mathematics. Mathematics should not be just the blind memorization of rules and formulas.

The Marvelous $15 Scientific Calculator

Virtually every student nowadays has a calculator. Some years ago, calculators were viewed by many math instructors as essentially cheating. But they are now almost universally accepted. Students need to understand "how" addition, subtraction, etc., work, but there is no need, in my opinion, to practice math operations just for speed.

The modern scientific calculator represents a triumph of technology. For $15 or less, you can buy a calculator that performs more than 200 mathematical functions, and with an accuracy of ± 1 in the 10th decimal place. Past mathematicians would have been astonished by these capabilities. These calculators let students carry out the calculations involved in basic mathematics, including differentiation and integration of simple equations, easily and quickly.

The Internet: Another Miraculous Aid

Another advantage that present day students and parents who are teaching math have is the Internet. Enter virtually any question about basic mathematics into an Internet "search engine", and you will receive dozens of relevant answers from numerous Web sites. Good Web sites include *Ask Dr. Math* (relatively informal), *MathWorld* (very comprehensive), *S.O.S. Math* (concise tutoring), *Paul's Online Math Notes* (lecture style), *Intmath* (easy reading) and *mathisfun* (great explanations). A freeware program for plotting mathematical functions is given as *www.mathgv.com.* Having access to a computer connected with the Internet is almost like having a personal tutor sitting at your side.

Chapter 1.

Some Importan [illegible] Ma [illegible]

In this initial [illegible] exam [illegible] buildi [illegible] of mathematics. These are the basic elements that [illegible] d in [illegible] tion [illegible] f equations. Being familiar with how these element [illegible] *ical* i [illegible] math [illegible] later sections, we will use the knowledge develop [illegible] view [illegible] ls th [illegible] algebra, trigonometry and calculus.

The Equal [illegible]

Many peopl [illegible] the d [illegible] is [illegible] werful symbol in the world. But they are wrong. Th [illegible] "equal" [illegible], is f [illegible] werful. The equal sign is the cornerstone of all p [illegible] hemati [illegible] the [illegible] s probably mankind's greatest invention.

When two t [illegible] hemati [illegible] ld say [illegible] own quantities") are equal, we show this *equality* b [illegible] **equ** [illegible] the [illegible] r of the equal sign lies in this one fact: if you do the [illegible] o eac [illegible] resultant equation is *still* equal. This is essentially t [illegible] algebr [illegible] mathe [illegible] olves around the imaginative use of this unique powe [illegible] a sign [illegible] o respect, and to rely on, the equal sign.

The equatio [illegible] simp [illegible]

$$[illegible] = x$$

But usually the equ [illegible] onsist [illegible] erms

$$[illegible] + B [illegible]$$

In mathema [illegible] s tha [illegible] hange [illegible] **variables**. They are usually written as lower ca [illegible] m near [illegible] the [illegible] e., *x*, *y* and *z*). The terms that do *not* change are call [illegible]. The [illegible] indi [illegible] tters from near the beginning of the alphabet (e.g., [illegible] a, b [illegible] he [illegible] ve, the variables, *x* and *z*, are *independent*, and c [illegible] val [illegible] ey a [illegible] he **independent variables**. But once the values of [illegible] ed, [illegible] y is [illegible] rmined, because A, B and C can not change. And *y* [illegible] the **dep** [illegible] **ria** [illegible] riable *y* is also frequently called the **unknown**, bec [illegible] value [illegible] se [illegible] rom the equation.

For student [illegible] ing m [illegible] the [illegible] be equal to several variables may seem strange. [illegible] is v [illegible] ney system. The dollars are equal (neglecting p [illegible] num [illegible] ters [illegible] nickels we have. These coins have a *fixed* value [illegible] ore c [illegible] he [illegible] ach coin we have is *variable*, and it is the value of th [illegible] nt va [illegible] ich [illegible] *u*, *v*, and *x*, that determines our total money. As an [illegible] wou [illegible]

$$\textit{total money (as ¢)} = Q(=25¢)u + D(=10¢)v + N(=5¢)x$$

Say we have 5 quarters, 3 dimes, and 4 nickels, then $u = 5$, $v = 3$ and $x = 4$. Introducing these values into the equation, we have

$$\begin{aligned} \textit{total money (as ¢)} &= Q(25¢) \times 5 + D(10¢) \times 3 + N(5¢) \times 4 \\ &= 125¢ + 30¢ + 20¢ \\ &= 175¢ \end{aligned}$$

Notice that we evaluated the expressions for the quarters, dimes and nickels *separately*, and then *added* these three values together to determine our total money. That is, we did just what the equation *told* us (in mathematical language) to do.

The total money here is given in cents. To convert the answer to dollars, we know that one dollar is equal to one hundred cents, and so one cent is *equal* to one dollar divided by 100. Moreover, one dollar divided by 100 cents is equal to *one*, and multiplying any term by one does not change its value. We can therefore convert *175¢* to *1.75$* (or *$1.75*) as shown in the last equation below:

$$\begin{aligned} \textit{one } \$ &= 100 \times \textit{one } ¢ \\ \textit{so} \quad \textit{one } ¢ &= \frac{\textit{one } \$}{100} \quad \textit{and also} \quad \frac{\textit{one } \$}{100\ ¢} = 1 \\ \textit{giving } 175\ ¢ &= 175\ ¢ \times \frac{\textit{one } \$}{100\ ¢} = 1.75\ \$ \end{aligned}$$

An important point to note here is that equations have **units** as well as numerical values. Also that these units can *cancel*, as in the last line of the equations above where the cent symbol, ¢, cancels because it appears as *both* numerator and denominator. Taking the account of the units that are involved in an equation is known as **dimensional analysis**. The units on each side of the equal sign must be the same (or consistent, as above), once the equation is solved. If this is not true, the equation or its solution *can not* be correct. Dimensional analysis is used extensively in chemistry, physics and engineering mathematics.

Equations are the essence of basic mathematics, and involve two steps. First, we must set up an equation that *accurately* represents the relationship between the independent variables and the unknown, and then we must be able to solve the equation to determine the value of that unknown.

The Numbers System

The first problem with math is the numbers. There are many types of numbers, and the student becomes confused about them. He finds it difficult to discern between "real numbers", "integers", "whole numbers", "natural numbers", etc. The Internet contains numerous articles on the nature, and the long history, of the classification of numbers. It also lists the generally accepted hierarchy of numbers, which we show in Table 1. Table 1 illustrates how the different types of numbers relate to one another.

Table 1

Hierarchy of Numbers

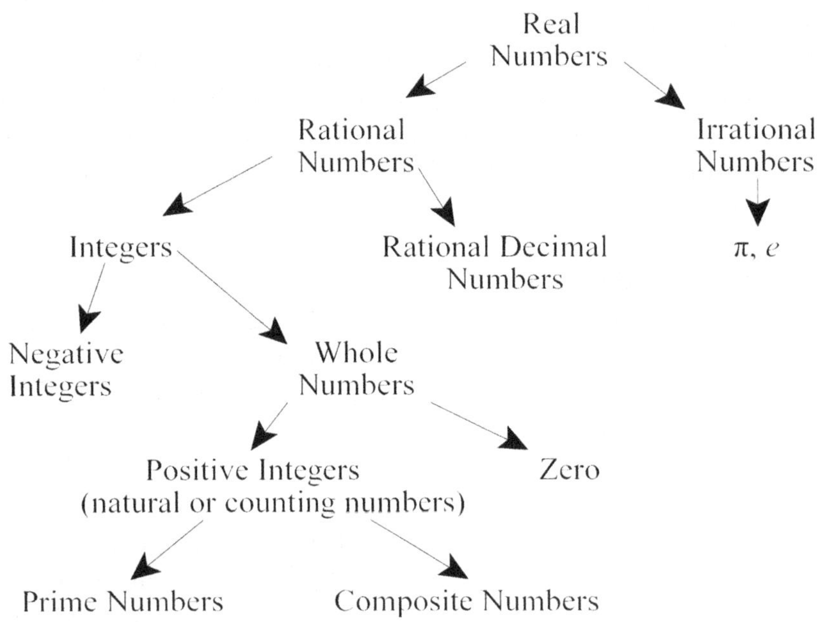

The numbers to which we refer in this book will be real numbers, unless otherwise specified. Basic mathematics is built on the **real number** set of ... -3, -2, -1, 0, 1, 2, 3... The real number set also includes **fractions** such as *1/2, 13/4, 5/6* and *1/64*. Fractions can be converted to **decimal numbers** by long division (or your calculator) giving, for the preceding fractions, the equivalent decimal values of *0.5, 3.25, 0.833...* and *0.015625*.

Real numbers are divided into **rational** and **irrational** numbers. Rational numbers can be produced by a *ratio* of integers (i.e., whole numbers). For example, the number *3* can be produced by taking the ratio of *6* to *2*, which is the same as *6* divided by *2* or *6/2*. Also, the denominator can be just *1*; that is, *-2/1* is equal to *-2*, *23443* is equal to *23443/1*, and therefore, *all* integers, positive or negative, are rational numbers. It is also possible for *decimal* numbers to be rational (a fact that students sometimes fail to recognize), if the decimal either terminates (i.e., has no leftover as in long division) or yields a repeating decimal sequence. For example, we have that the ratio of *1/64* = *0.015625*; that is, it divides out evenly without a leftover term, and so it terminates. It is thus a rational number. Other integer ratios, e.g., *1/3* = *0.3333...* or *10/11* = *0.909090...*, yield a repeating decimal ratio, and are also classified as rational numbers.

In contrast, irrational decimal numbers *do not* terminate or form repeating decimal sequences. Irrational numbers include $\sqrt{2}$ (see below), $\sqrt{3}$,$\sqrt{5}$ and others, as well as π (pi), the ratio of the circumference of a circle to its diameter, and *e*, the base for natural logarithms. The values of *π* and *e* have been calculated to *millions* of decimal places by computer without termination or any sign of a repeating decimal sequence being detected.

There are *four* numbers that cause trouble for beginning students — *e*, *π*, *0* and *i*. The letters *e* and π are just "short-hand" representations of two irrational numbers, *e* being equal to *2.7182...* and π equaling *3.1415...* . They occur in a large number of math equations, and with many different "looks". But they are *always* constants, and ultimately have just the numerical values shown.

The number, *0*, has special properties. We can add or subtract zero, and we can multiply by zero, but we can not *divide* by zero. This is a mistake in calculations often made by new math students. The quotient of *1/0*, if we consider *0* to be vanishingly small number, would be infinity. But infinity is an *undefined* number, so the mathematical operation of dividing by *0* is itself undefined and **not allowed**.

A second important fact about zero occurs when it is used as an **exponent**. When any number, positive or negative, is taken to the **zeroth power**, the value of the resultant number is one. In equation form, we would write, $(\pm \textit{any number})^0 = 1$. How any number taken to the zeroth power becomes equal to *1* is explained in the section on exponents (see below).

The letter *i* stands for the square root of *-1*, i.e., $i = \sqrt{-1}$. The **square root** of a number is a *second* number that, if multiplied by itself, will return the initial number. The square root sign is $\sqrt{}$. We say, for example, that $\sqrt{4} = 2$ because $2 \times 2 = 4$. But it is also true that $-2 \times -2 = 4$, because a negative number multiplied by another negative number yields a *positive* number. However, no number, whether negative or positive, can be multiplied by itself to produce a negative number. The square root of *-1*, written as $\sqrt{-1}$, is therefore undefined. To address this dilemma, mathematicians have defined $\sqrt{-1}$ as being ***imaginary***, and denoted by the symbol, *i*.

The imaginary unit, *i*, can be combined with real numbers to form **complex numbers**. Complex numbers have the form, $a + bi$, where *a* and *b* are real numbers. Despite containing an "imaginary" component, complex numbers are quite real (in the practical sense), and are used in applied mathematics and engineering. The topic of complex numbers is addressed later.

Degrees of Equality

Mathematics is not an exact science; i.e., it frequently yields answers that are not precisely exact, but which contain some degree of approximation (even if small). Mathematicians acknowledge this uncertainty by using a continuum of "equal" signs:

$y \equiv y$	*y* exactly equal *x*
$y = x$	*y* equal *x*, but with possibly some slight error
$y \approx x$	*y* approximately equal to *x*
$y \sim x$	*y* similar to, or loosely approximates *x*

$y \neq$ [illegible] not eq[illegible]
$y \geq$ [illegible] equal [illegible]er th[illegible]
$y \leq$ [illegible] equal [illegible] tha[illegible]

Various other equa[illegible] s[illegible]ch as [illegible] used [illegible] meaning is usually obvious from their appearance, a[illegible] of th[illegible]

Mathemati[illegible] [illegible]ons

Equations fi[illegible] [illegible]own to [illegible] ju[illegible] num[illegible] [illegible]st in lower mathematics). And we can do only *six* [illegible] [illegible]al op[illegible] [illegible]th nu[illegible] [illegible]umbers can only be added, subtracted, multipl[illegible] [illegible]ded, o[illegible] [illegible] divided using an **exponent**. Note that the numb[illegible] [illegible]ither k[illegible] [illegible]nkno[illegible] [illegible]pressed, e.g., by the variable, x) but the mathematica[illegible] [illegible]s are ju[illegible] [illegible]

Adding or [illegible] When [illegible] are [illegible]btracted, the operation is indicated by the + [illegible] [illegible]h as [illegible] *0.75* [illegible] The same symbols for adding and subtracting are [illegible] [illegible]braic [illegible] i.e., [illegible] *z, x + z - y*. When negative numbers are involv[illegible] [illegible]bolism b[illegible] [illegible]lig[illegible] [illegible]mplicated. We can *add* a negative number t[illegible] [illegible]umber [illegible] [illegible] is the same as simply subtracting *3*, or *2* [illegible] [illegible]ation[illegible] [illegible] ad[illegible] [illegible]tracted. Adding equations is straightforward:

$$(x + \ldots [illegible] q - v + \ldots [illegible] x \ldots [illegible] v + t$$

But if an equation i[illegible] [illegible] the *s*[illegible] [illegible] the [illegible] subtracted equation must be changed:

$$(x + \ldots [illegible] v + t) \ldots [illegible] -q \ldots$$

Multiplyin[illegible] [illegible]g. Mu[illegible] [illegible] di[illegible] [illegible]cated by *several* different symbols. Division [illegible] [illegible]n as [illegible] *x* d[illegible] [illegible] *y* as in an equation. Multiplication can [illegible] *a* × *b* [illegible] mply [illegible] [illegible]se of terms such as *Axy* to indicate multiplica[illegible] [illegible] con[illegible] [illegible] begi[illegible] [illegible]ents. They may subconsciously think of "*Axy*" as s[illegible] [illegible] *word* [illegible] [illegible] [illegible]et $A = 3, x = 2$ and $y = 4$, and then, substituting si[illegible] [illegible] wou[illegible] [illegible] = 3[illegible] *Ax4*. If we substitute all three values, then *Axy* = [illegible] which i[illegible] [illegible]

Equations c[illegible] [illegible]lied b[illegible] [illegible] o[illegible] [illegible] *equation*. For example, we can have *A(x + y + z)* [illegible] [illegible]al to *A*[illegible] [illegible]z. [illegible] [illegible]me number is the multiplier for an equation, every [illegible] [illegible]uatio[illegible] [illegible] mult[illegible] [illegible]at number. It can also be that we have one equation [illegible] [illegible]nother [illegible] *z)* [illegible] which is usually written just as *(x + y + z)(q + v +* [illegible]

We will illu[illegible] [illegible]tiplic[illegible] [illegible]uati[illegible] [illegible]ing with real numbers:

$$(2 + 3 + 2)(4 + 2 + 3) = ?$$
$$7 \times 9 = 63$$

but also we can calculate by

$$\begin{aligned} 2(4 + 2 + 3) &= 8 + 4 + 6 = 18 \\ 3(4 + 2 + 3) &= 12 + 6 + 9 = 27 \quad = 63 \\ 2(4 + 2 + 3) &= 8 + 4 + 6 = 18 \end{aligned}$$

To multiply equations, we simply multiply the second equation by *each* term in the first equation, and then sum the products.

If the equations involve unknown variables, the multiplication is still just the same, e.g.,

$$\begin{aligned} (x + y)(x + y) &= ? \\ x(x + y) &= xx + xy \\ y(x + y) &= yx + yy \quad = x^2 + 2xy + y^2 \end{aligned}$$

Notice that xx is given in *exponent* form, i.e., x^2 , in the final equation, and that $xy + yx$ is equal to $2xy$. Students sometimes fail to recognize that xy is equal to yx, but if we look at real numbers, all would agree that $3 \cdot 2 = 2 \cdot 3$.

Division of real numbers is done by traditional "long division", as known by most students. Algebraic long division is similar, and is the reverse of algebraic multiplication,

$$(x + 2)(2x^2 - 3x + 2) = ? \quad \textit{(multiplication)}$$

$$\begin{aligned} x(2x^2 - 3x + 2) &= 2x^3 - 3x^2 + 2x \\ + 2(2x^2 - 3x + 2) &= \qquad\quad 4x^2 - 6x + 4 \\ &= 2x^3 + x^2 - 4x + 4 \end{aligned}$$

And if we divide back, we obtain the original multiplicand of $3x^2 - 3x + 2$, and the answer comes out to be even.

$$\begin{array}{r|l} & 2x^2 - 3x + 2 \\ \hline x + 2 & 2x^3 + x^2 - 4x + 4 \\ & \underline{-(2x^3 + 4x^2)} \\ & \quad -3x^2 - 4x \\ & \quad \underline{-(-3x^2 - 6x)} \\ & \qquad\qquad 2x + 4 \\ & \qquad\qquad \underline{-(2x + 4)} \\ & \qquad\qquad\quad 0 \end{array} \quad \textit{(division)}$$

Students often ask, "But what if the answer in division doesn't come out to be even?" Then we have a **remainder**. With real numbers, we normally "take care" of the remainder by going to decimal numbers during the division process, e.g.,

$$
\begin{array}{r|l}
 & 4.714... \\
\hline
7 & 33.000... \\
 & \underline{28} \\
 & 5.0 \\
 & \underline{4.9} \\
 & 0.10 \\
 & \underline{0.07} \\
 & 0.030 \\
 & \underline{0.028} \\
 & 0.002...
\end{array}
\quad \textit{but also} \quad
\begin{array}{r|l}
 & 4 \\
\hline
7 & 33 \\
 & \underline{28} \\
 & 5/7 = \textit{remainder}
\end{array}
$$

$$\textit{and} \quad 4 + 5/7 \ (= 0.714...) = 4.714...$$

Using decimal division, we obtain a quotient of *4.714...* If we do "integer" division of *33* by *7*, our answer is *4* plus a remainder of *5/7*. But *5/7 = 0.714...*, and adding this remainder to the quotient *4* gives *4.714....* So when we add the remainder to the quotient, we find the same answer as obtained by decimal division.

This rule applies also for algebraic division. Consider this example,

$$
\begin{array}{r|l}
 & \phantom{3x^2 - {}} 3x - 8 \\
\hline
x + 2 & 3x^2 - 2x + 6 \\
 & \underline{-\ (3x^2 + 6x)} \\
 & -8x + 6 \\
 & \underline{-\ (-8x - 16)} \\
 & 22
\end{array}
$$

$$\textit{so} \ \frac{22}{x + 2} = \textit{remainder, and the answer} = 3x - 8 \ \textit{plus} \ \frac{22}{x + 2}$$

Note that the remainder is not *just* the number remaining, but the number remaining *divided* by the divisor. And note also that the answer is the quotient *plus* the remainder.

We can check the algebraic answer above by setting x equal to a small number such as *2*. Then $(x + 2) = 4$ and $(3x^2 - 2x + 6) = 14$, and *14* divided by *4* yields *3.5*. The algebraic answer consists of the quotient, $3x - 8$, plus the remainder, $22/(x + 2)$. For $x = 2$, the quotient becomes *-2*, and the remainder *22/4* or *5.5*. Adding *-2* to *5.5* yields *3.5,* which is equal to our preceding "arithmetic" answer of *3.5*. (n.b., This test works for all real numbers except $x = -2$, which causes the divisor to be zero, and division by zero is not allowed.)

Multiplication or division by exponents. The **exponent** indicates the number of things that are to be multiplied by each other. For example, $x^3 = xxx$, $2^4 = 2 \cdot 2 \cdot 2 \cdot 2 = 16$, $(x + 2)^3 = (x + 2)(x + 2)(x + 2)$, $A^3 = AAA$, etc. Note that the exponent indicates the number of *things* to be multiplied, and *not* the number of multiplication operations. That is, x^3 indicates there are three *x's* to be multiplied, but only two multiplications are involved, $x \cdot x \cdot x$. Students can be confused about this distinction. Also, it must be the *same* things that are being multiplied. If we multiply *AA* by *B*, the product *AAB* can not be described as A^3, since A^3 is defined as being $A \cdot A \cdot A$ (and also AAB would be different *numerically* from AAA).

Exponents are also used to indicate division. Let us consider that *xxx* is divided by *x*. In equation form, we would write,

$$\frac{xxx}{x} = xx\frac{x}{x} = xx; \quad or \quad \frac{x^3}{x} = x^{3-1} = x^2 = xx$$

$$equivalently, \ \frac{1}{x} = x^{-1}, \ and \ \frac{1}{x}\,xxx = x^3x^{-1} = x^{3-1} = x^2 = xx$$

Types of exponents. Exponents are used in *two* general forms. In the first form, mostly employed in **algebra**, the exponent is *known*, but the quantity being multiplied or divided by itself is unknown, e.g., x^3. In the second form, which relates mostly to **logarithms**, the quantity being multiplied or divided is known, but the exponent is *unknown*, e.g., 10^x.

In algebra, the basic *reciprocal* exponential operations involve **powers** and **roots**. Taking *x* to the third power is equivalent to multiplying three *x's* together, or x^3. Conversely, the third root of a number is the number which, when multiplied by itself two times, returns the original number. Some examples of usage are:

$$third\ power\ of\ x = x \cdot x \cdot x = x^3; \ third\ root\ of\ x^3 = \sqrt[3]{x^3} = x$$

$$third\ power\ of\ 2 = 2 \cdot 2 \cdot 2 = 8; \ third\ root\ of\ 8 = \sqrt[3]{8} = 2$$

for algebraic expressions

$$fourth\ power\ of\ (x^2 + 2) = (x^2 + 2)^4; \ fourth\ root\ (x^2 + 2)^4 = \sqrt[4]{(x^2 + 2)^4} = (x^2 + 2)$$

The second root, or *square root*, is usually indicated just by √, with the descriptive *2* being omitted. It is also common to refer to the third power as "..as cubed or being cubed", and the third root as being the "cube root."

The *nth* root can also be written as an exponent without the √ root sign being involved. We can have, e.g.,

$\sqrt{2x+2}$ ... $2)^{\frac{1}{2}}$ ... $2)^{0.5}$

$\sqrt[4]{2x+2}$ ... $2)^{\frac{1}{4}}$ ... $2)^{0.25}$

$\sqrt[5]{2x+2}$ ... $2)^{\frac{1}{5}}$ ... $2)^{0.20}$

If we should ... ion which has been raised to the *fifth power*, say, (2x ... ould ...

$\sqrt[5]{(2x}$ ... $+ 2)^1 = (2x + 2)$

The laws of ... the exponent laws are sometimes given ... al numbers, and *m* and *n* to be *positive integers.*" ... *m* and *n*) to be "positive integers" can misle ... to th ... *not* be fractional or decimal, or negative or zero. ... true ...

(a^m) ... $= 2^{2+3} = 32$

$2^1 = 2$

$4^2 = 16$

(a^m) ... $64 = 2^6$

$(a$ ... $(ab)^n$ ... $= 6^2 = 36$

Logarithm ... ents ... me way as do roots and powers. One of the most im ... erties ... of *roots* and *powers*, and of *logarithms* and *exp* ... ns. For instance, let $y = 4$, then

$$y = 4$$

taking sq. *root* *of each side gives,* $\sqrt{y} = \sqrt{4}$ *(sides remain equal)*

and, *squaring* *each side gives,* $y^2 = 4^2$ *(sides remain equal)*

This is a fairly intuitive finding. Taking each side to *any* root, or raising each side to *any* power does not change the equality. With logarithms and exponents, this property is not so readily visualized, but it is nonetheless true. For example,

$$y = 4$$

then *taking the* log *of each side gives,* $\log y = \log 4 = 0.6020...$ *(sides stay equal)*

and making each side the *exponent* *gives* $10^y = 10^4 = 10{,}000$ *(sides stay equal)*

Much of the difficulty arises because many students do not have a good understanding of logarithms. The **common logarithm**, based on the number *10*, is defined as:

$$number = 10^{\log}$$

Note that the log term is just an *exponent*. If the **log** is *2*, then we have

$$number = 10^2 = 10 \times 10 = 100$$

The defining equation also allows us to find the **antilogarithm** of the log term, which is the number *produced* by the given log term. The **antilog** of *2*, for example, is *100.*

This works well when the log terms (or *logs*) are integers. But what if the logs are decimal numbers? Suppose that we have

$$number = 10^{0.3010...}$$

where the log is the irrational number, *0.3010...* Taking *10* to the *0.3010...* power on your calculator gives an answer of *2* (essentially). The log of *2* is therefore *0.3010...*, and the relationship is usually written in the form, $\log_{(10)} 2 = 0.3010...$ This says that the logarithm of *2* to *base 10* is equal to *0.3010...*

A second type of logarithm, the **natural logarithm**, is based on *e*, and is defined as

$$number = e^{\ln}$$

The natural log is abbreviated as **ln** to distinguish it from the common base (10) log which is denoted by **log**. It is also be expressed, using *2* as an example, as $ln_{(e)}\,2 = 0.6931...$ The symbol *e* has a mysterious look, but it is only "shorthand" for an irrational number which can be approximated as *2.71828...* We could write *e* as *2.71828...*, but it wouldn't be as convenient. It happens, however, that *e* is the **natural base** for analyzing growth processes (see section on **compound interest** in Chap. 3 for a derivation of *e*), and thus is useful in many areas of science. The mathematics of common vs. natural logs is *just the same*, except for the value of the base.

Each number has a specific logarithm. In past days, logarithms of individual numbers had to be looked up in a "log" table. Modern scientific calculators can find both the common and natural logarithms of any real number, and calculate the "power value" of these logarithms (i.e., as 10^{log} and e^{ln}) easily and accurately. They also warn of errors. Try to take the logarithm of a negative number and the calculator will return, "Math ERROR".

A common problem in physical chemistry and other sciences is to convert between natural and common logarithms. Before calculators, one had to remember the conversion factor of *2.303*, which was applied as *log given number = 2.303 ln given number*. This conversion factor can be shown found by comparing the calculator values for *log 10* and *ln 10*. i.e.,

$$\log 10 = 1, \quad so \quad 10 = 10^1$$
$$and \ \ln 10 = 2.30258..., \quad so \quad 10 = e^{2.30258...}$$

The exponent for the base *e* must therefore be *2.30258...* (*~2.303*) times larger than the exponent to base *10* to return the number *10*. This ratio between common and natural logarithms holds for any real number, as calculator testing will show.

Being exponents, logarithms are governed by the Laws of Exponents. For multiplication, Law 1 says $(a^n)(a^m) = a^{n+m}$, where *a* is any real number. With common logarithms, *a* is the *base* of *10*, and *n* and *m* are *log(1)* and *log(2)*, respectively. As an example,

$$by\ arithmetic,\ 2.5 \times 3.1 = 7.75$$

$$by\ logs = 10^{0.3979}(\log 2.5) \times 10^{0.4914}(\log 3.1) = 10^{(0.3979+0.4914=0.8893)}$$

$$10^{0.8893} = 7.7499, \quad or \ \log 0.8893 = antilog\ 7.7499$$

For division, Law 2 states that $a^n/a^m = a^{n-m}$, or in logarithmic terms,

$$\frac{6.7}{2.8} = 2.3928,\ but\ also = 0.8260...(\log 6.7) - 0.4471...(\log 2.8) = 0.3789...$$
$$and\ the\ antilog\ of\ 0.3789... = 2.3927$$

Logarithms can also be used to calculate *powers* and *roots* in a convenient manner, e.g.,

$$x^3 = x \cdot x \cdot x = \log(x) + \log(x) + \log(x) = 3 \log(x)$$

$$2^3 = 3 \log(2) = 3(0.3010...) = 0.903...; \ \textit{antilog} = 7.998...(\sim 8)$$

$$\sqrt[3]{8} = \frac{\log(8)}{3} = \frac{0.903}{3} = 0.301; \ \textit{antilog} = 1.999.... \ (\sim 2)$$

As a separate point, note that the calculator can yield logarithms which at first glance may appear strange. But these can be explained as follows. For instance,

$$\log 0.056 = -1.2518...$$

$$\textit{this } \log \textit{ comes from} \quad 0.056 = \frac{5.6}{100} = \frac{0.7481...(\log 5.6)}{2 \ (\log 100)} = 0.7481 - 2 = -1.2518...$$

Similarly, one can have

$$\log 1367 = 3.1357...$$

$$\textit{from} \quad 1367 = 1.367 \times 1000, \ \textit{so} \ 0.1357...(\log 1.367) + 3 \ (\log 1000) = 3.1357...$$

These logs cause no problems, because if they are entered as the exponents of *10* in the calculator, they will return the correct answers, i.e., $10^{-1.2518...} = 0.056$, and $10^{3.1357...} = 1367$.

Math textbooks often list natural logarithms as having these two properties:

$$1) \quad \ln e^x = x$$
$$\textit{and } 2) \quad e^{\ln x} = x$$

These can be confusing because of the "*e* factor", but they are understandable if we look at them to the base *10* (all logarithms must have the same properties). Property 1) is the equivalent of saying that the log of 10^2 (i.e., *100*) is *2*, which we know to be true. Property 2) says that if, say, $x = 100$, then $10^{\log 100 \text{ (which = 2)}} = 100$. This is the same as saying 10^2 is equal to *100*.

Finally, logarithms are decimal numbers which we use as exponents. Students may wonder how a number can be multiplied by itself some *fractional* number of times. This question is explained in the section below on **Positive fractional exponents**.

The Slide [illegible] ntifi [illegible]

Before the [illegible] ientifi [illegible] lations [illegible] using a **slide rule**. The slide rule is simply a *mechan* [illegible] er mul [illegible] div [illegible] ers by adding or subtracting logarithms. It consi [illegible] low [illegible] nd a [illegible] per ruler. Both rulers are marked off with the numbe [illegible] *0* at [illegible] hat co [illegible] o their logarithms. This is illustrated in Table [illegible] cal slid [illegible] ve m [illegible] ecimal subdivisions.)

Table 2

[illegible] struct [illegible] 10-[illegible] Rule

[illegible]	[illegible] of [illegible]	[illegible] ition in Inches [illegible] 10" Scale
[illegible]	[illegible]	[illegible] 0
[illegible]	[illegible]	[illegible] 010
[illegible]	[illegible]	[illegible] 771
[illegible]	[illegible]	[illegible] 021
[illegible]	[illegible]	[illegible] 990
[illegible]	[illegible]	[illegible] 782
[illegible]	[illegible]	[illegible] 451
[illegible]	[illegible]	[illegible] 031
[illegible]	[illegible]	[illegible] 542
[illegible]	[illegible]	[illegible] 000

To multiply [illegible] the zer [illegible] upper [illegible] *2* on the lower scale, and read the number on the [illegible] which [illegible] he [illegible] er scale. This number is *6*. We have mechanically [illegible] g for [illegible] the [illegible] *771)* which gives *7.782*, the log of *6*. To divide, re [illegible] cedur [illegible] *2* o [illegible] scale over *6* on the lower scale, and read the numbe [illegible] er sca [illegible] belo [illegible] that log *1* = *0*) on the upper scale. This subtrac [illegible] n *7.78* [illegible] s 4. [illegible] for *3*. Multiplication and division are thus ea [illegible] out on [illegible].

However, sli [illegible] acco [illegible] nly [illegible] between *0* and *10*. To deal with large numbers, the [illegible] **enti** [illegible] wa [illegible]. In scientific notation, a number such as *65,* [illegible] as [illegible] *009* [illegible] nential notation as 6.5343×10^4 (but a slide rule ca [illegible] to [illegible] in [illegible] alculations, the numbers are truncated, i.e., *6.53* [illegible] *6.53)* [illegible] mpl [illegible] cation using scientific notation,

$$(7,[illegible])(6,23[illegible]) = [illegible] \times 10^{[illegible]} [illegible] 10^4)(6.23 \times 10^3)$$

$$= [illegible](1.2[illegible]) [illegible] 10^3 \times 10^4 \times 10^3$$

$$= [illegible](1.2[illegible]) [illegible] 10^{3+4+3}$$

$$= [illegible] \times 10^{1[illegible]}$$

$$[illegible] \times 10^{1[illegible]}$$

Scientific notation calculations make use of both Law 1, $(a^n)(a^m) = a^{n+m}$, and Law 2, $a^n/a^m = a^{n-m}$ of the Laws of Exponents. For the case of division,

$$\frac{4{,}340}{(1{,}200)(48{,}700)} = \frac{4.34 \ x\ 10^3}{1.2 \ x\ 4.87 \ x\ 10^3 \ x\ 10^4}$$

$$= \frac{4.34}{(1.2)(4.87)} \ x\ 10^3 \ x\ 10^{-3} \ x\ 10^{-4}$$

$$= 0.74 \ x\ 10^{3-3-4}$$

$$= 0.74 \ x\ 10^{-4}$$

$$= 7.4 \ x\ 10^{-5}$$

Allowable Numbers in Exponents

Students often wonder if *all* of the different types of real numbers can be used as exponents. In this section, we examine that question by taking each type of number in turn.

Positive integers. When the exponent n is a positive integer in expressions such as x^n, A^n, $(x + y)^n$, *etc.*, it signifies the *number* of things that are to be *multiplied* by each other. There are no restrictions on n, except that as a positive integer it must be equal to, or larger, than *1*.

$$2^3 = 2 \bullet 2 \bullet 2 = 8$$

$$(x + y)^3 = (x + y)(x + y)(x + y)$$

Negative integers. When the exponent is a negative integer, it indicates the number of things by which the expression is to be divided, or the number of *reciprocals* by which it is to be multiplied. It must be an integer of *-1* or smaller (i.e., *-2, -3, etc.*).

$$2^{-3} = \frac{1}{2} \bullet \frac{1}{2} \bullet \frac{1}{2} = \frac{1}{8}$$

$$(x + y)^{-3} = \frac{1}{(x + y)} \bullet \frac{1}{(x + y)} \bullet \frac{1}{(x + y)}$$

Positive fractional exponent. Can we have a fractional exponent such as *2/3*? Yes, this is possible. We can take, for example, the cubic root of a number that has been squared, i.e.,

$$\sqrt[3]{2^2} = \sqrt[3]{4} = 1.587\ldots \quad (\textit{directly by calculator})$$

but also

$$\sqrt[3]{4} = 4^{\frac{1}{3}} = (2^2)^{\frac{1}{3}} = (2)^{2\bullet\frac{1}{3}} = 2^{\frac{2}{3}} = 2^{0.666\ldots} = 1.587\ldots \quad (\textit{by exponent})$$

And we can treat *any* fractional exponent in this way. For example,

$$2^{\frac{11}{38}} = \sqrt[38]{2^{11}} = \sqrt[38]{2048} = 1.222\ldots$$

$$\textit{but also } 2^{\frac{11}{38}} = 2^{0.2894\ldots} = 1.222\ldots$$

That is, we can either take the *38th root* of the number we obtain by raising *2* to its *11th power*, or we can convert the fractional exponent to a decimal exponent, and then apply that decimal number as the exponent of the base. Either procedure is easily done with a scientific calculator.

Negative fractional exponent. A negative exponent changes the expression to a *reciprocal*. For example, x^{-2} is simply equal to $1/x^2$. This is not changed if the exponent is a fractional exponent which we convert to a decimal number. Take $2^{-1/4}$,

$$2^{-\frac{1}{4}} = \frac{1}{2^{\frac{1}{4}}} = \frac{1}{2^{0.25}} = \frac{1}{1.1892\ldots} = 0.8408\ldots$$

and also,

$$(0.8408\ldots)^{-4} = 2.000\ldots, \text{ "}\textit{reversing}\text{" } \textit{the reciprocal}$$

Fractional negative exponents therefore behave in the same way as negative integer exponents.

Exponents of zero. Any number raised to the *zeroth* power, which is to say having *zero* as an exponent, is equal to *1*. This is true whether the number is positive or negative, or an integer or fraction. This is shown by,

$$\frac{(any\ number)^x}{(any\ number)^x} = 1,$$

but also

$$\frac{(any\ number)^x}{(any\ number)^x} = (any\ number)^{x-x};\ and\ so\ (any\ number)^0 = 1$$

All numbers, whether positive or negative, integer or fraction, or even zero, can therefore function as exponents. Exponents are not limited to being just "positive integers."

Not allowed numbers. However, the *numbers upon which exponents act* do have limitations. The logarithm of a *negative* number can not be found because no logarithm can yield a negative number. Similarly, the even roots of negative numbers can not be found because no negative number, multiplied by itself, can produce a negative number. A negative number multiplied by a negative number yields a *positive* number.

Zeroth power of zero. Finally there is the question of the value of the zeroth power of zero, or 0^0. From the equation above, we see that $0^0 = 0/0$. And since generally, division by zero is not allowed, 0^0 might be thought to be indeterminate. But other considerations in mathematics suggest, in fact, that $0^0 = 1$, and this is now the majority consensus among mathematicians. The question of 0^0, which has been debated for hundreds of years, is discussed in *Ask Dr. Math*, and other mathematics Web sites.

Factorial numbers. Later in this book, we will encounter factorials, i.e., *1!*, *2!*, *3!*, *4!*, and so on. The factorial is shorthand notation for a particular type of mathematical operation, namely,

$$\begin{aligned} one\ factorial &= 1! = 1 \\ two\ factorial &= 2! = 1\times2 = 2 \\ three\ factorial &= 3! = 1\times2\times3 = 6 \\ four\ factorial &= 4! = 1\times2\times3\times4 = 24 \\ &\cdots \end{aligned}$$

This brings up the question then of what is the value of *0!* ? It turns out that for mathematical consistency (just as with 0^0), the value of *0!* is defined as being *1*, or

$$zero\ factorial = 0! = 1$$

Closing no[illegible]

Beginning [illegible] fear [illegible] ma[illegible] [illegible]lve a virtually limitless number of different types o[illegible] [illegible]al op[illegible] [illegible]is i[illegible] We introduce several areas of mathematics in this [illegible] [illegible]ding [illegible] calc[illegible] [illegible] each case, once the equations are interpreted and [illegible] [illegible]rm, th[illegible] [illegible]ved [illegible] [illegible]e *six* mathematical operations we have discussed [illegible] [illegible]ition, [illegible] [illegible] and division, and multiplication and division by mea[illegible] [illegible]ents.

Chapter 2.

Equations and Functions

Up this point, we have spoken only of equations, or conditions of equality. But mathematicians tend to speak more about **functions**, especially in calculus. Beginning students are often confused as to the difference between an equation and a function. In this Chapter, we will look at the relationship between equations and functions, beginning first with equations.

The Power of Equations

The power of an equation lies primarily in this one fact. If we do the *same thing* to each side of the equation, the equation remains *equal*. This simple property of the equal sign is used extensively in algebra, and is important in the solving of equations in general.

The "things" that we can do to an equation consist essentially of the *mathematical operations* described in Chapter 1. That is, we can add or subtract, multiply or divide, or do exponential multiplications and divisions on each side of the equation. Each of these types of operations are illustrated below.

1) We can *add or subtract* the same known (or unknown) quantity to each side, and the equation stays equal. Consider the equation

$$y = 2x + 4$$

If we subtract *4* from each side, we have

$$y - 4 = 2x + (4 - 4 = 0), \quad or\ y - 4 = 2x$$

Alternatively, we could subtract *2x* from each side giving

$$y - 2x = (2x - 2x = 0) + 4, \quad or\ y - 2x = 4$$

In this way, we can "move" any term from one side of an equation to the other by simply adding (or subtracting) that term from *each* side. Students are sometimes told to "just take the term to the other side (of the equal sign) and change its sign." This works, but it does not explain to the student *why* this movement is justified mathematically.

Adding or subtracting can also be used to "reduce" complex equations. As a simple example, let

$$6x + 3 = 2x + 15$$

If we subtract *2x* and *3* from each side of the equation, we obtain the reduced equation,

$$(6x - 2x) + (3 - 3) = (2x - 2x) + (15 - 3)$$
$$\textit{or,}\quad 4x = 12$$
$$x = 3$$

2) We can also *multiply or divide* each side of the equation by the same factor, and the equation will remain equal. For a multi-term equation such as below, dividing each side by *4* simplifies the equation, but maintains the equality,

$$4y = 6x + 2z + 4$$
$$\textit{dividing by 4,}\quad \frac{4}{4}y = \frac{1}{4}(6x + 2z + 4)$$
$$\textit{yields,}\quad y = 1.5x + 0.5z + 1$$

Notice that *each* term on both sides of the equation must be divided (or multiplied) by the factor involved (in this case, *4*).

In other instances, we may be able to simplify the equation by multiplication such as in this example where multiplying each side by *6* leads to a solution of the equation:

$$\textit{For,}\quad \frac{2x - 3}{2} + \frac{3x}{6} = \frac{4x + 4}{3},\ \textit{multiplying each side by 6 gives}$$

$$\frac{6}{2}(2x - 3) + \frac{6}{6}(3x) = \frac{6}{3}(4x + 4),\ \textit{or}\quad 3(2x - 3) + 1(3x) = 2(4x + 4)$$

$$\textit{and then,}\quad 6x - 9 + 3x = 8x + 8,\ \textit{or}\ (9 - 8)x = 8 + 9,$$

$$\textit{and}\quad x = 17$$

3) and finally, we can manipulate the equation by the use of *exponents*. Most students are at ease with adding/subtracting and multiplying/dividing from grade school mathematics. The reciprocal mathematical operations of powers/roots and logarithms/antilogarithms are not so clear to them, however. But all exponential calculations are governed by the *Laws of Exponents* (pg. 10) and, using these, we can (with practice) learn to apply exponential methods in solving equations. Here are a few examples of exponential calculations with equations:

1)

$$10^x = 49$$
$$\log(10^x) = \log(49)$$
$$x \log(10) = \log(49);\quad \textit{and,}\ \log(10) = 1,$$
$$\textit{so}\ x = \log(49) = 1.690...$$

2)

$$\ln(x - 2) + \ln(2x - 3) = 2\ln(x)$$
$$\ln((x - 2)(2x - 3)) = \ln(x^2)$$
$$\textit{therefore}\ \ (x - 2)(2x - 3) = x^2,$$
$$\textit{and}\ \ 2x^2 - 7x + 6 = x^2,$$
$$\textit{or}\ \ x^2 - 7x + 6 = 0$$
$$\textit{which (by inspection) has solutions of } x = 1, \textit{ and } x = 6$$

3)

$$\sqrt[4]{x - 3} = 2$$
$$(\sqrt[4]{x - 3})^4 = 2^4,$$
$$\textit{or}\ \ x - 3 = 16;\ \textit{and}\ x = 19$$

4)

$$\sqrt{x} + \sqrt{x + 4} = 1$$
$$\sqrt{x + 4} = 1 - \sqrt{x}$$
$$x + 4 = (1 - \sqrt{x})^2 = 1 - 2\sqrt{x} + x$$
$$4 = 1 - 2\sqrt{x}$$
$$\sqrt{x} = -\frac{3}{2} = -1.5,$$
$$\textit{and}\ \ x = (-1.5)^2 = 2.25$$

One common mistake that inexperienced students make with exponents involves parentheses, e.g.,

$$(x + 4)^2 \textit{ is } \underline{\textit{not}} \textit{ equal } (x^2 + 4^2)$$

$$\textit{let } x = 3;\ \ \textit{then } (3^2 + 4^2) = 25;\ \ \textit{but should be } (3 + 4)^2 = 49$$

The expression, "..$(x + 4)^2$..", means to multiply $(x + 4)$ by itself, i.e., $(x + 4) \times (x + 4)$ which is equal to $x^2 + 8x + 16$ by algebraic multiplication. If $x = 3$, the value of this expanded expression is *49*, just as it is for the lower "arithmetic solution." One can not simply take the exponent "inside the parentheses" and apply it to each term individually.

This *peculiarity* of exponent calculation is incorporated in a rule in mathematics known as the **order of operations**, which specifies the order in which math operations *must* be done in solving equations in order to achieve the correct answer.. Consider this equation, assuming that we know initially that $x = 2$,

$= 3(x$ [illegible] *and let* $x = 2$

then clear pa[illegible] $= 3($[illegible] 6 [illegible] $= 3{\cdot}5^2 + 6 - (12 \div 4)$

m[illegible] $= 3{\cdot}5$[illegible] 12 [illegible] $+ 6 - (12 \div 4)$

di[illegible] $= 75 +$ [illegible] $6 - 3$

a[illegible] $= 75$ [illegible] $= 81$

su[illegible] $=$ [illegible]

This order of opera[illegible] the righ[illegible] But [illegible] mistakenly *first* multiplied the units inside the init[illegible] coe[illegible] we [illegible] $6 + 9)^2 = 225$, and not the true value of *75* which [illegible] $3(2 +$ [illegible] Also, note that if there are sets of *nested* parenthes[illegible] ses mu[illegible] ed f[illegible] *rmost* parentheses outward.

In example[illegible] ove, [illegible] *roo*[illegible] which are also known as **radicals**. Radical[illegible] phas[illegible] ays, [illegible] can be easily converted to exponential notatio[illegible]

$\sqrt[4]{\ldots}$ [illegible]

Problems involvin[illegible] n be sol[illegible] your [illegible] using either the *nth root* or *exponential* key fun[illegible]

Adding and [illegible] g "wh[illegible] uations.

In addition [illegible] ng e[illegible] *sing*[illegible] n, we can add, or subtract, *whole* equations. Conside[illegible] ons, y[illegible] and [illegible] These are classified as **linear algebraic equatio**[illegible] ey in[illegible] aic[illegible] nd, when plotted, yield a *straight* line. The property [illegible] to add [illegible] plies to *all* equations, however, and is not limited t[illegible] quati[illegible] we [illegible] simplicity.)

Since y_1 is [illegible] *4*, an[illegible] *6*, we [illegible] *equal things* to each side of the equal sign when we [illegible] equa[illegible] ether.

[illegible]

We can verify the addition here by taking different values of x. Let $x = 2$, for example, and then $y_1 = 8$ and $y_2 = 0$, so $y_1 + y_2 = 8 + 0$, *or* 8. But also $y_1 + y_2 = y_{total} = 5x - 2$ which, for $x = 2$, gives just the same answer of 8.

The equations can also be subtracted. Using the same two equations, we find that

$$\begin{aligned} y_1 &= 2x + 4 \\ -y_2 &= -(3x - 6) \\ \hline y_1 - y_2 = y_{total} &= -x + 10 \end{aligned}$$

Assume that $x = 3$. Then, $y_1 = 6 + 4 = 10$, and $-y_2 = -9 + 6 = -3$, with $y_1 - y_2 = 10 - 3 = 7$. And if we substitute $x = 3$ into the equation $y_{total} = -x + 10$, we obtain the same answer of 7.

Addition or subtraction of equations is therefore a valid mathematical operation. And this is true not just for two equations, but even if an *unlimited* number of equations are added or subtracted.

If we plot any two linear algebraic equations, and the plotted lines are not parallel, they *must* intersect. (Note: we can detect parallel lines without plotting because they will either have the same equation, be multiples of the same equation, or be displaced from one another by a constant.) At the point of intersection, y_1 is equal to y_2, which gives

$$2x + 4 \ (= y_1) = 3x - 6 \ (= y_2)$$
$$\textit{or, } x = 10$$

$$\textit{then, } y_1 = 2(10) + 4 = 24$$
$$\textit{and, } y_2 = 3(10) - 6 = 24$$

The point of intersection for the two lines is therefore $x = 10$, $y = 24$. Finding the point of intersection of two lines is a common problem in algebraic linear equations.

Substitution in equations.

In some cases, we may have equations involving two variables such as $2x + 4y = 12$ and $3x - 2y = 10$. A system of equations of this type involves **"two equations and two unknowns."** And we can solve these equations by **substitution** where we solve the second equation for y in terms of x, and then substitute this value for y in the first equation. As an example,

$$2x + 4y = 12$$
$$3x - 2y = 10$$

$$\textit{from the 2nd equation, } y = \frac{3x - 10}{2}$$

$$\textit{and } 2x + 4\left(\frac{3x - 10}{2}\right) = 12$$

$$\textit{giving } 2x + 6x - 20 = 12, \textit{ or } 8x = 32;\ x = 4$$

$$\textit{then } 2(4) + 4y = 12; \textit{ or } 4y = 4;\ y = 1$$

This is a powerful means for solving algebraic equations, and it is applicable anytime that "the number of *independent* equations" is equal to "the number of unknowns."

Independent equations are satisfied by just *one* pair of numbers such as $x = 4, y = 1$ in the case above. Equations such as $2x + y = 4$, and $2x + y = 3$ are called *inconsistent*, because no pair of numbers will satisfy both equations. Equations such as $x + y = 4$ and $2x + 2y = 8$, which can be shown by algebra to derive from the same equation, are called *dependent* equations. Equations which describe parallel lines, e.g., $y = x + 1$ and $y = x + 2$, have no point of intersection, and can not be solved by any pair of x,y values. Independent equations have one pair of x,y values that satisfies both equations *simultaneously*, and so independent equations are also commonly referred to as **simultaneous equations**.

Solving simultaneous equations involving *three* or more variables is more difficult, and is usually undertaken using **matrix algebra** (see below).

Conversion of units.

The variables involved in practical mathematics have *units*. For example, if we travel for *4 hours* at a speed of *10 miles per hour*, the distance we cover will be *40 miles*. The distance is not just *40* but *40 miles*. If it is necessary to know the distance in *kilometers*, the miles must be converted to kilometers. The calculations are,

$$4 \textit{ hrs } x \frac{10 \textit{ miles}}{hr} = 40 \textit{ miles} \quad (\textit{hrs cancel})$$

$$\textit{and } {\sim}0.62 \textit{ miles} = 1 \textit{ km};\quad \textit{so } 1 = \frac{1 \textit{ km}}{{\sim}0.62 \textit{ miles}} \quad (\textit{divide sides by } 0.62 \textit{ miles})$$

$$\textit{then } 4 \textit{ hrs } x \frac{10 \textit{ miles}}{hr} x\ (1) = 40 \textit{ miles } x \left(\frac{1 \textit{ km}}{{\sim}0.62 \textit{ miles}} = 1\right) = {\sim}64.5 \textit{ km}$$

Notice that we have multiplied the left side of the last equation by *1*, and the right side by the conversion ratio. We are thus mathematically consistent; that is, we have multiplied both sides of the equation by *equal* values.

The properties of the equal sign are used extensively in algebra, as we will see in a later section in this book on algebra.

A Function of ...

The dictionary defines **function** as "...a mathematical correspondence that assigns *exactly* (my italics) one element of one set to each element of the same or another set." Some math textbooks invoke the concept of an undefined "function machine." The *independent* variables are put into the machine, and the correct values for the *dependent* variables come out. The machine converts independent variables into dependent variables "...by following certain sets of rules"; the word *equation* is avoided. The **domain** is defined as the *set* of independent variables (which can be limited) that can be put into the machine, and the **range** is the *set* of dependent variables that result.

I believe this definition of function is unnecessarily sophisticated for beginning students. Virtually every function the student will encounter is based upon an equation. In my opinion, it will not handicap students, in the early part of their learning, to relate functions with equations. However, while functions may be based upon an equations, *not all* equations yield functions. The **critical concept** for a function is that it can have *only* one value of y for each value of x. Many equations yield *two* or more values of y for each x. These equations can not act as functions.

The function sign.

Consider an equation such as $y = 4x + 3$. The value of y is not totally fixed by x. For instance, if $x = 0$, then y still equals 3. But, the value of y changes as a *function* of x. We say that, ".. y is a function of x," and denote the fact by use of the **function sign**, f, written as,

$$y = f(x)$$

Note that, since $y = f(x)$ and also $y = 4x + 3$, it must follow that $f(x) = 4x + 3$. We can therefore write $f(x)$ as a "shorthand" for the right side of the equation, which may be known or *unknown*. The notation, $f(x)$, tells us that y varies with x, but we can not say *how* without knowing the specific equation ($4x + 3$ here) that $f(x)$ represents.

Keep in mind that f(x) is unique, and relates to *just one* certain equation between y and x. If another equation for y becomes involved (say, $y = 3x + 1$), we must denote this by using a different letter in the function sign, e.g., $y = g(x)$. If the dependent variable, y, is the function of two independent variables, such as $y = 2x + 3z$, then we write $y = f(x,z)$. For three independent variables, we would have $y = f(x,z,u)$, and so on. The expression, $f(a)$, is also often seen. It means, "the value of y when $x = a$ in the equation under study," or $y_{(x = a)} = f(a)$.

The function is frequently not written as just $f(x)$. In manufacturing, for example, the profit is a *function of* both the costs of labor and of material. And this dependency may be emphasized by writing the function as $P(l,m)$. But it is always necessary to have an equation that accurately relates the cost of labor and material to profit in order to evaluate the function.

Algebraic ca [illegible] an also [illegible] nctions, such as $f(x) + g(x)$, $f(x) - g(x)$, $f(x)g(x)$, [illegible]. If we [illegible] know [illegible] ations that the various functions represent, we can c[illegible] calcul[illegible] $f(x)$ [illegible] and $g(x) = 2x - 4$, then we have,

$f(x) +$ [illegible] $+ 2)$ [illegible] $2 - 4 = 5x - 2$

$f(x) -$ [illegible] $+ 2$ [illegible] $4) =$ [illegible] $2 + 4 = x + 6$

[illegible] $= ($[illegible] $8x - 8$

[illegible]

In more advanced [illegible] he diff[illegible] functio[illegible] arried through several stages of calculation before b[illegible] solve[illegible].

The Functi[illegible] lus

One of the [illegible] s th[illegible] t h[illegible] ions occurs in calculus. We will briefly discuss [illegible], and i[illegible] ated [illegible], here as a *pro tem* introduction to calculus itself.

Calculus ha[illegible] ial part[illegible]

1) The **func**[illegible] as $f($[illegible] an e[illegible] t describes the mathematical relationship of the [illegible] ariab[illegible] *depe*[illegible] le.

2) The **deri**[illegible] en as [illegible] is a[illegible] n (i.e., another equation) or *constant* that descr[illegible] t wh[illegible] ende[illegible] changes with respect to the independent variab[illegible] n the d[illegible] y dif[illegible] on of the function.

3) The **anti**[illegible] which [illegible] by "[illegible] erentiating" the *derivative, f '(x)*, to yield its "origina[illegible] *f(x)*, a[illegible]

4) the **integ**[illegible] as *F(*[illegible] eff[illegible] *anti-derivative* of the original function, $f(x)$. We [illegible] egral [illegible] b a[illegible] nd obtain $F(b) - F(a)$, which is the area under the c[illegible] ted by t[illegible] *on f(*[illegible] $x = b$ and $x = a$.

It is importa[illegible] mind [illegible] *diff*[illegible] s that are used, these being **F(x)**, **f(x)** and **f '(x)**, to r[illegible] *integ*[illegible] *ific*[illegible] *rivative*, respectively.

This list has [illegible] ms, b[illegible] res[illegible] of calculus. Be familiar with these four terms, a[illegible] a tre[illegible] d in[illegible] ing calculus.

Differentia[illegible]

In calculus, [illegible] h op[illegible] is differ[illegible] Differentiation can be thought of as "finding the *diff*[illegible] esults [illegible] nden[illegible], when the independent variable, x, is chang[illegible] ery [illegible] nt."

Differentiat[illegible] y intro[illegible] *phic*[illegible] spect to a plotted curve that is generated by a cert[illegible]

Consider the function

$$y = x^2, \quad where\ f(x) = x^2$$

When the function, $y = x^2$, is plotted, it yields the curve in Fig. 2-1. We have added two straight lines. The solid straight line is the *tangent* to the curve at point *a*, and the dotted straight line is a *secant line* that connects the points *a* and *b* on the curve. (n.b., a tangent line *touches* a curve at just one point, while a secant line intercepts the curve at *two* points.)

The important point to notice in Fig. 2-1 is that, as *b* moves down the curve toward *a*, the slope of the secant approaches the slope of the tangent at *a*. Moreover, as *b* approaches *a*, the distance, *h*, between the *x* intercepts of points *b* and *a* approaches zero.

The slope of a straight line is set by

$$slope = \frac{\Delta y}{\Delta x} = \frac{y_2 - y_1}{x_2 - x_1}$$

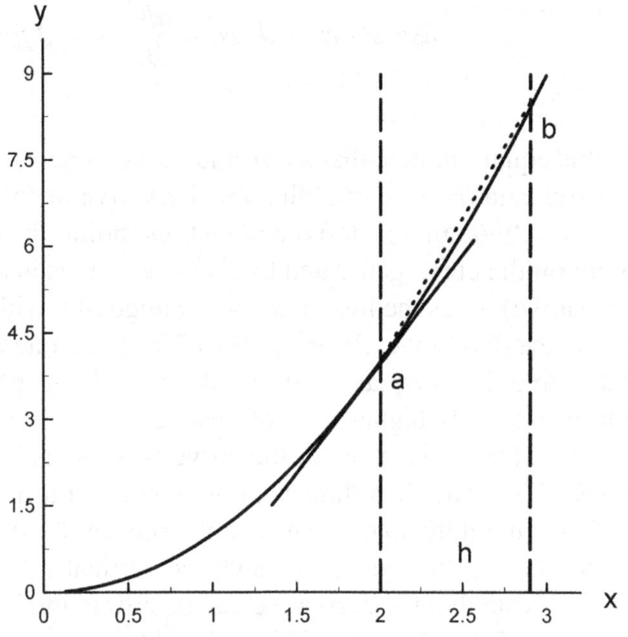

Fig. 2-1 Differentiation of y = x^2

Since $y = x^2 = f(x)$, $y_2 = f(x_2) = f(x + h)$, and $y_1 = f(x_1) = f(x)$. And, the values for x are simply $x_2 = x + h$ and $x_1 = x$.

The slope of the secant (which becomes the slope of the tangent) can therefore be calculated according to

$$slope\ of\ secant = \frac{y_2 - y_1}{x_2 - x_1} = \frac{f(x_2) - f(x_1)}{x_2 - x_1} = \frac{f(x + h) - f(x)}{(x + h) - x} = \frac{f(x + h) - f(x)}{h}$$

$$since\ f(x) = x^2,\ we\ then\ have\ \frac{(x + h)^2 - x^2}{h} = \frac{(x^2 + 2xh + h^2) - x^2}{h} = \frac{h(2x + h)}{h}$$

$$slope\ of\ secant\ (or\ tangent) = 2x + h \rightarrow 2x,\ as\ h \rightarrow 0$$

$$slope\ of\ tangent = 2x = f'(x) = (the\ \underline{derivative})$$

The slope of the tangent at the point, *a*, on the curve is therefore equal to $2x$. This is the *instantaneous* rate of change of y with respect to x at this point. It is also the **derivative** of the function, $y = x^2$, at the point *a* on the curve. This can be mathematically summarized as,

$$\frac{\Delta y}{\Delta x} \rightarrow \frac{dy}{dx} = \frac{d\,f(x)}{dx} = f'(x) = \frac{d\,(x^2)}{dx} = 2x$$

$$\textit{also shortened as,} \quad \frac{dy}{dx}\ (\textit{the \underline{differential}}) = 2x\ (\textit{the \underline{derivative}})$$

In this equation, note that as Δy and Δx become very small, they go to dy and dx. Also recall that $y = f(x)$, and $f(x) = x^2$, and that the derivative of $f(x)$ is written as $f\,'(x)$.

Although it is derived at just *one* point, the derivative gives the slope of the curve at *any point* on the curve generated by $f(x) = x^2$. This is important to remember. Also the slope (or derivative) gives the instantaneous change of y with respect to x at that point. For instance, at $x = 1$, the derivative, $2x$, is equal to 2, and the rate of change is *2 y/each x*. At $x = 5$, the derivative, $2x$, is equal to 10, and the rate of change is *10 y/each x*. Larger values of x give correspondingly higher rates of change.

At $x=0$, the slope of the curve is *zero*. This is allowable, but the slope can never be *vertical* (i.e., the slope line can not go straight up). That would correspond to a slope of *infinity*, which is not definable. (This is one criterion for determining a "differentiable" function; the slope to its plotted curve can never be vertical.)

Note that the derivative can be a function itself. The derivative of $2x$ from above is a function of x, and can be differentiated to give a second derivative of 2. We can see this graphically. If we plot, $y = 2x$, it yields a straight line through $0,0$ with a constant slope of 2. The derivative of $f(x)$ is written as $f\,'(x)$, and the derivative of $f\,'(x)$ is written as $f\,''(x)$. This signifies that it is the second derivative of the original function, $f(x)$. If further derivatives are taken, they are written $f\,'''(x)$ and so on.

The differentiating of a function $f(x)$ to yield a series of derivatives may also be indicated by:

$$f'(x) = \frac{d}{dx}f(x) \quad \textit{and also } f''(x) = \frac{d^2}{dx^2}f(x), \textit{ and so on}$$

$$\textit{also since } y = f(x),\ \ y' = \frac{d}{dx}y = \frac{dy}{dx},\ \ \textit{just as } f'(x) = \frac{d}{dx}f(x)$$

It is also possible to "turn around" the differentiation, and convert the derivative back to the original function, which is now called the **anti-derivative**. In the present case, $f\,'(x) = 2x$, which as we know, reverts to $f(x) = y = x^2$, if we reverse the differentiation process. Note that *f '(x)* gives the *slope* of the curve at point *a*, while *f(x)* gives the *value* of *y* at point *a*. If $x = 10$, for example, the slope is $2x$ is $2(10)$ or *20 y/each x*, while the value of *y* is x^2 or $10^2 = 100$.

Integration.

Consider a straight line that stretches from *0* to *3* along the *x* axis. If we divide it into two, the length of each segment is *3/2* and the sum of the two segments is $3/2 + 3/2 = 3$. If we divide the line into an infinite number of segments, the length of the line is equal to the summation of this infinite number of segments, and is *still* equal to *3*. As an equation, we would write,

$$length\ of\ line = \sum_{n=1}^{n=\infty} dx_n = 3$$

This equation says the length of the line is equal to the summation (Σ symbol) of all of the segments from segment (1) to segment (n=∞); and moreover that the summation of this endless number of segments is just equal to *3*. We can accept this intuitively.

This process can also be cast as an **integration** ("..to make many into one"), which is written as

$$length\ of\ line = \int_0^3 dx\ (the\ integral) \Rightarrow |x|_0^3 = (x{=}3) - (x{=}0) = 3\ (solved\ integral)$$

The integral is marked by the **integral symbol,** $\int$, where the subscript *0* and the superscript *3* give the **limits of integration**. Essentially what we do is to integrate the expression (see Chap. 9 on calculus), and then evaluate it as the difference between the upper and lower limits of integration, i.e.,

$$(expression) \int dx = x + C\ (integral, = F(x))$$

$$upper\ limit\ integral\ (F(3)) - lower\ limit\ integral\ (F(0)) = value\ of\ integral$$

Now consider the function, $y = x^2$, which generates the **curve** plotted in Fig. 2-1. The area under this curve can be considered to be made up of an infinite number of *rectangles*, which when integrated, are equal to the area under the curve. The area of each micro-rectangle is given by the height of the rectangle, *y*, times its width, *dx*. What is the value of *y*? It is a function of *x*, and is given at *every point* on the curve by $y = x^2$. We can therefore substitute x^2 for *y*, and the area of each micro-rectangle is then, $micro\text{-}area = x^2 \times dx$. To find the total area under the curve from $x = 0$ to $x = 3$, we must integrate all of the micro-rectangles, which amounts to solving the integral,

$$\int_0^3 x^2 [illegible]$$

To find the integra[illegible] must fi[illegible] which [illegible]rentiated, yields x^2. (Integration is the reverse of dif[illegible]) Fro[illegible] of di[illegible]on (see Chap. 9 on calculus), we know that

$$[illegible] = 3[illegible]$$

[illegible] *whi[illegible]* [illegible] $\frac{1}{3}x^3 = F(x)$

And therefore we h[illegible]

$$\int_0^3 x^2 d[illegible] \; [illegible] \; \left. \frac{1}{3}[illegible] \right| [illegible] = 9 [illegible] = \textit{area under curve}$$

In general t[illegible] ation [illegible] as

$$\int_a^b [illegible] \; [illegible] F(b) [illegible]$$

which is the **funda[illegible] ion** [illegible].

This equati[illegible] essari[illegible] Th[illegible] ve even the most infinitesimal dx segment of x must [illegible] pe. [illegible] an [illegible] e of y so accurately represent the height of the micro[illegible] t $y \times$ [illegible] e t[illegible] e micro-rectangle? This problem, along with [illegible] about th[illegible] aning[illegible], caused mathematicians to consider calculus a[illegible] en"fo[illegible] year[illegible] introduction by Newton and Leibniz (independe[illegible] 5.

It was not u[illegible] 54 th[illegible] mann [illegible]) proved by means of the **Riemann sum** that [illegible] e mi[illegible] les [illegible] the definite integral between a and b under the cur[illegible] l by th[illegible] His [illegible]mmarized in the equation,

$$\lim_{n \to \infty} [illegible] \Delta x [illegible]$$

This indicates that as the number of "micro-rectangles" becomes infinitely large, the summation of their areas becomes equal to the integral function, regardless of where the "height", given by $f(x_k^*)$, may be located within the subinterval, Δx. In the function $f(x_k^*)$, the subscript, k, indicates that the function involves a particular k micro-rectangle of the infinite series of micro-rectangles. The superscript, *, denotes a single value of y (as calculated from $f(x_k^*)$) that occurs within the Δx subinterval in question.

Closing note

As we go later to the study of calculus, keep in mind from this Chapter that:

$f(x) = y = x^2$ generates the curve in Fig. 2-1, and gives the *value* of y for each point of x on the curve,

$f'(x) = y' = 2x$ is the *derivative* of $f(x)$ and indicates the *instantaneous rate of change* in y with x at each point on the curve, and

$F(x) = f(x)dx$ = *the anti-derivative of* $f(x)$ which is the *integral* function that integrates the *area* under the curve.

Chapter 3.

Rate of Doing

Much of practical mathematics is concerned with the **rate of doing** something. This can include such things as the rate of transferring a material, the speed (i.e., rate of travel) of a vehicle or projectile, the rate of percent returns on money, or the rate of natural growth and decay. In this Chapter, we will study examples of the calculations that are encountered in determining "the rate of doing."

Rate of Material Transport

One example of material transport is when you fill the gasoline tank on your automobile. This takes a few minutes, and it is described by the equation,

$$gals\ (in\ tank) = \frac{gals}{min}\ x\ min$$

Note that the rate is "gallons *per* minute." The term *per* means that the number of gallons is *divided* by minutes. Also notice that if we know *two* of the terms in the equation, we can always determine the *third*. For example, if a particular gasoline pump has a flow rate of 2.5 gals/min, we can calculate how many minutes will be required to fill our 20-gallon tank by re-arranging the equation as,

$$20\text{-}gal\ tank\ x\ \frac{\text{min}}{2.5\ gals} = \text{min}$$

$$\frac{20\ gals}{2.5\ gals}\ x\ \text{min} = 8\ \text{min}$$

We can also have *competing* rates of flow. Let us assume that a large fuel storage tank is transferring fuel to two delivery systems at rates of 10 gals/min and 5 gals/min, respectively. If at the same time, we are adding more fuel into the storage tank at a rate of 100 gals/min, the *net* rate at which fuel is accumulating in the storage tank is given by

$$net\ rate\ of\ fuel\ added = \left(\frac{100\ gals}{min} - \frac{10\ gals}{min} - \frac{5\ gals}{min}\right) = \frac{85\ gals}{min}$$

To determine the net rate of material transport, we simply sum the various flow rates that are involved, taking into account whether the flow is into a particular vessel or system (positive), or *out* of the vessel or system (negative).

Rate of Travel (or Velocity)

Velocity is the time rate of doing travel in a *linear* manner in a particular direction. In the US, velocity is commonly given in miles *per* hour (mi/hr) or feet *per* second (ft/sec). It can be an *average* velocity, as in "..we traveled 100 miles in 2 hrs total time for an average of 50 miles per hour," or an *instantaneous* velocity, as in "..right now, we are doing 50 miles per hour."

We describe the rate of travel by the equation

$$\textit{distance (miles)} = \frac{\textit{miles}}{\textit{hour}} \times \textit{time (hours)}, \quad \textit{e.g.,} \quad \frac{50\ \textit{miles}}{\textit{hour}} \times 2\ \textit{hours} = 100\ \textit{miles traveled}$$

The distance can be expressed in any units of distance (miles, feet, kilometers, etc.), and the time in any units of time (hours, minutes, seconds, etc.). And if we know any two terms in the equation, we can always determine the third. If we travel *100* miles in *2* hours, our rate of travel is *50 miles/hr*. And to travel *100* miles at a rate of *50 miles/hr* will require *2 hours*. And so on.

This equation is used in the old "...two trains are traveling" problems. The essential ideas in solving these problems is to note that 1) a 20 mph train can never catch a 40 mph train unless it gets some sort of head start, and 2) if the trains are headed toward each other, they are closing at 60 mph (20 mph + 40 mph).

Acceleration/deceleration. In real life, the velocity does not remain constant. It will increase or decrease as the result of a **force** applied with, or *against*, the direction of the vehicle or projectile. The relationship between force and acceleration is given by

$$F = ma; \textit{ where } F = \textit{force}, \ m = \textit{mass}, \textit{ and } a = \textit{acceleration}$$

In textbook math problems, the force specified is often the force of gravity because it is constant, and easy to define. The force of gravity is given by

$$F = = mg; \textit{ where } g \textit{ is the acceleration resulting from gravity}$$

On earth, the acceleration, *g*, resulting from the force of gravity is *32.17 ft/sec²*, or *~32 ft/sec²*. The units of acceleration appear strange, but they come from this consideration,

$$\textit{velocity} = \frac{\textit{ft}}{\text{sec}}$$

$$\textit{acceleration} = \textit{change in velocity } \underline{\textit{per}} \text{ sec} = \frac{\textit{ft}}{\text{sec}} \times \frac{1}{\text{sec}} = \frac{\textit{ft}}{\text{sec}^2} \textit{ or } \textit{ft}\text{-sec}^{-2}$$

As an exam [illegible] ation [illegible] ulate [illegible] ce/velocity relationships of a ball thrown *straight up*. [illegible] vels [illegible] the [illegible] rs of this equation,

$$distan \ldots \text{[illegible]} \ldots (\frac{ft}{\text{sec}})\cdot t(\text{sec}) + 5\ ft$$

The distance the ba [illegible] h up [illegible] own [illegible] on of time, and it is indicated by *s(t)*. The first term [illegible] repre [illegible] celer [illegible] d by gravity, the second term is the initial *velocity* [illegible] nd the [illegible] s the [illegible] *oint* (i.e., *5 ft* above the earth's surface). Note that [illegible] s intro [illegible] fir [illegible] three terms are *ft*, so the equation is dimens [illegible] ct; *ft* [illegible] = *ft* [illegible]

The ball lea [illegible] at *5* [illegible] an i [illegible] ity of *128 ft/sec*, but it begins to be *immediately* slo [illegible] ty. T [illegible] igh [illegible] ld be in, say *3* seconds, we introduce *t = 3 sec* [illegible] ation, [illegible]

$$distance\ in\ ft \ldots = s(t = \text{[illegible]} \ldots + 128\ (\frac{ft}{\text{sec}})\cdot 3(\text{sec}) + 5\ ft$$

$$s(t = 3) = \text{[illegible]}\ 128 \ldots = \ldots 384\ ft + 5\ ft = 245\ ft$$

The ball would thus [illegible] the ai [illegible] It would have been at *389 ft* if there was no effect [illegible]

One might [illegible] coeffi [illegible] e ac [illegible] rm should be *-32*, and not -16. But we are looking [illegible] *age* v [illegible] e ba [illegible] *neous* velocity at the end of the *2 sec* is *2× -32 ft/se* [illegible], bu [illegible] y at [illegible] s *0*. Since the rate of acceleration is *line* [illegible] we di [illegible] al v [illegible] *64 ft/sec* by *2*, which gives the average velocity of [illegible] Note t [illegible] ks f [illegible] d of time. The average velocity for a ball falling fo [illegible] examp [illegible] 2 o [illegible] Therefore we use *-16 ft/sec* in the acceleration ter [illegible] al for [illegible] the [illegible] travel.

If we differ [illegible] nction [illegible] espe [illegible] Chap. 9 on calculus), we obtain its first derivative, [illegible] gives th [illegible] *neo* [illegible] f the ball at each point in its travel.

$$\ldots = -3 \ldots (\frac{ft}{\text{sec}})$$

We can use *s'(t)* to [illegible] highe [illegible] at th [illegible] each in its flight. At the apex of its travel, the ball's [illegible] omes [illegible] *ro*. [illegible] ing *s'(t) = 0*, we can find the *time* at which the ba [illegible] becom [illegible] o ze [illegible]

$$s'(t) = -32t + 128 = 0$$

$$then\ t = \frac{-128}{-32} = 4\ seconds$$

Then by substituting *t = 4* into the original function *s(t)*, we can calculate that the height of the ball at its apex is *261* feet.

Finally, we can differentiate the first derivative to obtain the *second derivative*, which is simply $s''(t) = -32\ ft/sec^2$. This is the acceleration that results from the force of gravity, and it is applied, in the direction toward the earth, to the ball at all times and heights in its travel. It is a constant, and is independent of time or height.

When the ball starts to fall back to earth, it will accelerate at a constant *-32 ft/sec*, and after *4 sec*, the ball will have returned to earth, with its velocity at that point being *-128 ft/sec*. This is just the reverse of the ball's initial upward velocity of *128 ft/sec* (The fact that the two velocities are identical, except for direction, is a manifestation of the "conservation of energy" from thermodynamics.)

The definition of acceleration as the increase in velocity *per sec* may lead some students to think "...you must deal in *whole* seconds." This is not true. An object that has been under an acceleration of $32\ ft/sec^2$ for *3.786* seconds will have a velocity of $32\ ft/sec^2 \times 3.786\ sec$, or *121.15 ft/sec*, at the end of that time.

This discussion of the flight of a ball illustrates the power of math. Watch a ball be thrown upward and then fall back to the earth, and you can say little except perhaps to guess at its maximum height. But if you know the ball's initial velocity, you can employ the function *s(t)*, along with its first and second derivatives *s'(t)* and *s''(t)*, to *calculate* the total distance the ball travels, its highest point, and its velocity at each point of travel. These simple points are the historic beginning of the mathematics that guides our space exploration programs.

Automotive velocities. Similar calculations can help us quantify automobile performance. A modern drag racer can travel a quarter mile (*1320 ft*) in *4.5 sec*, with a timed terminal speed of *330 mph*. So we have

$$1320\ ft = \frac{a}{2}t^2 + 0\ (no\ initial\ speed) + 0\ (starting\ at\ 0\ ft)$$

$$so\ \ average\ acceleration = a = \frac{1320 \times 2}{4.5^2} = 132\ ft\text{-sec}^{-2}$$

$$or\ in\ terms\ of\ g,\ \ \frac{132\ ft\text{-sec}^{-2}}{32\ ft\text{-sec}^{-2}} = 4.1\ g$$

But note that the terminal speed calculated using this average rate of acceleration is *405 mph*, which is *not correct*, and is substantially higher than the measured velocity of *330 mph*, i.e.,

$$acceleration\ \frac{ft}{\text{sec}^2} \times time\ (\text{sec}) = \frac{ft}{\text{sec}}\ velocity$$

$$so\ acceleration = 132\ ft\text{-sec}^{-2} \times 4.5\ \text{sec} = 594\ ft\text{-sec}^{-1}\ (terminal\ velocity)$$

$$mph = \frac{60\ mph}{88\ ft\text{-sec}^{-1}} \times 594\ ft\text{-sec}^{-1} = 405\ mph$$

The error is that we have assumed that the car accelerates at a constant rate (i.e., the calculated average rate), but in fact the acceleration rate is variable over the quarter mile, and is greater first and then lessens near the terminal end as the wind resistance increases.

The g-force loading in a car crash is also of automotive interest. A car hitting a solid wall loses all of its velocity within the length of the crumble zone, or about *3 ft*. Assume that the car is traveling at *60 mph*, and that it loses velocity linearly over the *3 ft* interval. This gives an average velocity during the impact of *30 mph* or $44\ ft\text{-}sec^{-1}$ We have then that

$$time\ involved\ in\ crash = \frac{3\ ft}{44\ ft\text{-sec}^{-1}} = 0.068\ \text{sec}$$

$$then\ 3\ ft = \frac{a\ ft\text{-sec}^{-2}}{2}(0.068\ \text{sec})^2,\ so\ a = \frac{3\ ft \times 2}{(0.068\ \text{sec})^2} = 1298\ ft\text{-sec}^{-2} = 40.5\ g\text{-}force$$

The driver is thus subjected to an average force of *40.5 g*, or for a *200 lb* man, a weight of *8000 lbs* driving him forward.

Rate of Money Earned via Interest

When money is borrowed or loaned, a fee is paid for the use of that money which is called the **interest**. Borrow *$100* for a year at an *8%* interest *rate*, and you must pay back *$108* at year's end. Percent comes from the Latin term *per centum* which means "per hundred."

To make percent calculations, we divide the percent number (or percent "rate") by *100*, and then multiply, or divide, the money borrowed (called the **principal**) by the resultant decimal number. As an equation, we would write

$$Principal\ x\ \frac{Percent\ rate}{100} = Interest$$

This single equation is *all* that is involved in percentage calculations. If you know two of the three terms, then you can determine the third by simple algebra.

To show this, let the principal be \$150, and the percent rate 6%. We then have

$$\$150 \times \frac{6\%}{100} = \$9 \text{ interest earned}$$

Given the interest earned, and the principal, find the interest rate,

$$\text{interest rate} = \frac{\$9}{\$150} \times 100 = 6\%$$

And, finally, find the principal involved when the interest earned is \$9, and the percent rate is 6%.

$$\frac{\$9}{\frac{6\%}{100}} = \frac{\$9}{6\%} \times 100 = \$150 \text{ principal}$$

Compound Interest

Interest involves a *time period.* Loan *\$100* to a bank at a *6%* Annual Percentage Rate (APR) and, at the end of one year, you will receive your *\$100* back plus *\$6* interest money. You then have *\$106*, which is equal to your original *\$100 x 1.06.* Leave the total *\$106* in the bank for another year, and you will accrue *(\$100 x 1.06) x 1.06*, which is equal to $\$100 \times (1.06)^2$. Let the money *compound annually* for 10 years, and the final amount will be $\$100 \times (1.06)^{10}$, or \$179.08.

The formula for determining the money accrued as it is being compounded *annually* is therefore

$$D = D_0(1 + r)^t$$

where D is the total deposit at the end of time, t; D_0 is the original deposit; r is the percent rate *as the decimal fraction;* and t is the number of years invested.

Banks often offer *quarterly compounding* or even *continuous compounding.* For an annual interest rate of, say, 8%, the *quarterly interest rate* is 8% divided by 4. But the compounding occurs four times more frequently per year. The equation for annual compounding is therefore modified to

$$D = D_0(1 + \frac{r}{n})^{nt}$$

where D, D_0, r, and [illegible] ame m[illegible] and n is [illegible] er of compoundings *per year*. For a deposit of $10[illegible] ompound[illegible] times p[illegible] total money accumulated in one year is

$$\textit{[illegible] money} \ [illegible] \ 1 + [illegible] \quad \$108.24$$

This is \$0.24 more [illegible] that [illegible] be[illegible] ated at an 8% APR. Assume the interest was compo[illegible] ntinuously [illegible] sing [illegible] tor, we approximate as 10,000 times per year). Th[illegible] ey in thi[illegible] ould [illegible] 09 more,

$$\textit{[illegible] money} = [illegible] + [illegible] = \$108.33$$

If your money is ac[illegible] or a peri[illegible] e, [illegible] the difference between annual and continuous com[illegible] ecom[illegible] rec[illegible]

$$\textit{Annual [illegible] unding} \ [illegible] .08)^{20} = \$466.10$$

$$\textit{Con[illegible] Compou[illegible]} \ [illegible] 00(1 [illegible])^{10000x20} = \$495.30$$

Leonard Eu[illegible] *783)*, [illegible] great[illegible] icians of all time, studied the mathematics of com[illegible] est and [illegible] is gr[illegible] found the following relationship,

$$[illegible] \left(1 + \frac{1}{x}\right)^{x} \ [illegible] \textit{const} [illegible] 82818...$$

The term, "lim (as) [illegible] ns the [illegible] alue [illegible] ression would reach if x becomes infinitely [illegible] days, [illegible] air[illegible] oximation (*2.7182805...*) of the value of e by simpl[illegible] g x as [illegible] on [illegible]. However, in 1748, Euler evaluated e by the [illegible] **ion** b[illegible] otain[illegible] for *e* that is accurate to *18 decimal places*.

$$e = 1 [illegible] \frac{1}{2!} + [illegible] \ldots$$

How the value of th[illegible] *(1* [illegible] be ob[illegible] a series equation is explained later in the book in [illegible] **n seri**[illegible] tions.

As we cons[illegible] ever, w[illegible] ri[illegible] he natural logarithm system is defined as *number* [illegible] we se[illegible] for th[illegible] equal to the interest rate × time in years (i.e., *0.08* × 2[illegible] he sam[illegible] for [illegible] ed *($495.30)* as our original continuous compou[illegible] ion:

Continuous Compounding: $\$100\ (1 + \frac{0.08}{10000})^{(10000\ x\ 20)} = \$495.30$

but <u>also</u> $\$100\ e^{(0.08\ x\ 20)} = \$495.30$

This means that:

$$e^{0.08} = (1 + \frac{0.08}{10000})^{10000}$$

This identity can be derived as follows, where *r* is the growth (or interest) rate given as the *decimal* (i.e., *8%* is given as *0.08*), *N* is the number of compoundings per year, and *t* is the number of years:

$$for\ (1 + \frac{r}{N})^{N\,t},\ let\ \frac{r}{N} = \frac{1}{n},\ giving\ N = nr$$

$$then,\ (1 + \frac{1}{n})^{nrt} = (1 + \frac{1}{n})^{n})^{rt} = e^{rt}$$

Note that this derivation uses *Law 4 of Exponents* which says that a^{nm} is equal to $(a^n)^m$, or in this case $(1 + 1/n)^{nrt}$ is equal to $((1+1/n)^n)^{rt}$, which is equal to e^{rt}.

The equation $D = D_o\, e^{(r\,x\,t)}$ is a very convenient way for calculating continuously compounded growth. Note that, although *t* is in years, it does not have to be in *whole* years.

$$D = \$100\ e^{(0.08\ x\ 3.45)} = \$100\ x\ 1.318.. = \$131.80$$

Also we can calculate interest to base *10* by setting e^{rt} to be equal to $10^{rt/2.303}$; that is,

$$D = \$100 \times 10^{\frac{.08\ x\ 3.45}{2.303}} = \$131.78$$

The power of compound interest can be illustrated by the story of Manhattan Island. Manhattan Island was purchased from the Indians by the Dutch in 1626 for *60* guilders (approximately *$24*). Manhattan is currently estimated to be worth of the order of *$250* billion. However, if the Indians had invested the *$24* at *6.25%* interest (a not unusual industrial bond interest rate), the worth of that investment in 2007 would be

$$worth\ of\ investment = \$24(e^{(.0625\ x\ 381)}) = \$527\ billion$$

If the Indians had been able to invest at 6.25% interest, they would have gotten considerably the better of the deal. But interestingly, if their interest rate was just 6.0%, the worth of their investment would be only *\$203* billion, or *\$324* billion *less*. Even 0.25% makes a huge difference, if the period of investment is long enough.

Rate of Growth

In the continuous compounding of money, as each bit of interest is earned, it begins earning interest itself. This is the way in which **growth** occurs in nature. Therefore we can use the exponential equation above to calculate rates of growth. Since different symbols are used, the equation may look slightly different. For example, if we were studying bacteria, we might set $\mathbf{B_0}$ as the amount of bacteria at the starting time *($t = 0$)*, $\mathbf{B_t}$ as the amount of bacteria after time, t, and **k** as the **growth constant** (equivalent to the interest rate in the compounding of money). This gives

$$B_t = B_0 e^{kt}$$

This equation has four "unknowns". But we need to know *only* the time at which the amount of bacteria has doubled (as determined by some physical measurement) to calculate the growth constant. Let the doubling time be *5 hrs*, then we have

$$\frac{B_t}{B_0} = 2 = e^{kt}$$

$$\textit{taking } \ln \textit{ each side, } \ln 2 = \ln(e^{kt}), \textit{ yields } 0.693 = kt$$

$$\textit{then, } 0.693... = k \times t\ (=5\,hrs);\ \textit{so}\ \ k = \frac{0.693...}{5\ hrs} = \frac{0.138...}{hr}$$

This gives the growth constant, k, as 0.138.../hr. Note that the units of k are reciprocal time. Multiplying k by time yields a unit-less number (which is required since exponents have no units). Once we know k, we can calculate the *relative* change in bacteria for any time. And if either B_0 or B_t are known, the other can be quantitatively calculated.

Decay is the opposite of growth. And it can be treated by the same exponential growth equation, but in "reverse." A common problem is treating the radiation decay of radioactive elements. Assume that the radiation decreases by *one half* over some time period. This defines the **half life** of the radioactive element. Say the half life is *5 hrs* in the present case, which gives

$$\frac{R_t}{R_0} = 0.5 = e^{-\gamma t}$$

$$\ln(0.5) = -0.693... = -\gamma t;\ \textit{then}\ \ \gamma = \frac{0.693...}{5\ hrs} = 0.138...$$

Notice that the equation and treatment are just the same as for growth, except that the growth constant is replaced by the **decay constant**, γ, and the exponent $-\gamma t$ is *negative*.

The half life is very useful in science. Given the half life for any substance, one can easily calculate the time required for that substance to be reduced to 1/2, 1/4, 1/8...of its original value.

Chapter 4.

Plotting of Equa[illegible]

Math textbo[illegible] conta[illegible] numb[illegible]s. There is a reason for this. By plotting out an equa[illegible]ph, yo[illegible]kly [illegible] equation behaves. Few of us can look at an equa[illegible]aliz[illegible]pens[illegible]endent variable(s) takes on different values. Al[illegible]son be[illegible]ferent [illegible]quations is made especially easy. In this Chapter, we [illegible] graph[illegible]from di[illegible]uations, and point out the information that th[illegible]

A downloa[illegible]-frie[illegible]re [illegible]plotting mathematical functions is currently availab[illegible]he gra[illegible]of its [illegible]www.mathgv.com.*

Linear Eq[illegible]

Linear equa[illegible]*straig*[illegible]y are classified as *first-degree* **polynomials** (see b[illegible]finition[illegible]mial[illegible]he independent variable, x, is raised to the first p[illegible] $y =$[illegible] is [illegible] second power, as in $y = x^2 + 6$, the equation is a *se*[illegible] polyn[illegible]he e[illegible]tains x^3, it is a *third-degree* polynomial, and so [illegible]-degre[illegible]-deg[illegible]mials yield *curves* when plotted.

As an exam[illegible] equa[illegible]ider [illegible] ways of selling gasoline. In one case, we charg[illegible]en, bu[illegible]our[illegible] $3. In the other case, there is no pump-use fee, but t[illegible]is $3 p[illegible] W[illegible]nd the "best-buy" by plotting the two equations, [illegible]

The two str[illegible]ntersec[illegible] gallons of gas, and [illegible]he $3[illegible] price is the less ex[illegible] more [illegible] gallons of gas are p[illegible]ote th[illegible] can always find th[illegible] sim[illegible] looking at the *lowe*[illegible]graph[illegible]

The genera[illegible]near equation is

$$y = mx + [illegible]$$

where m is the slop[illegible]and b [illegible] constant. The valu[illegible]wn as [illegible] "y-intercept" beca[illegible] 0, th[illegible] linear graph interce[illegible]s at th[illegible] value b. We can al[illegible] "x-intercept" which is [illegible] wh[illegible] 0. From the equati[illegible] $y = m$[illegible] we see that when y [illegible]al to [illegible]

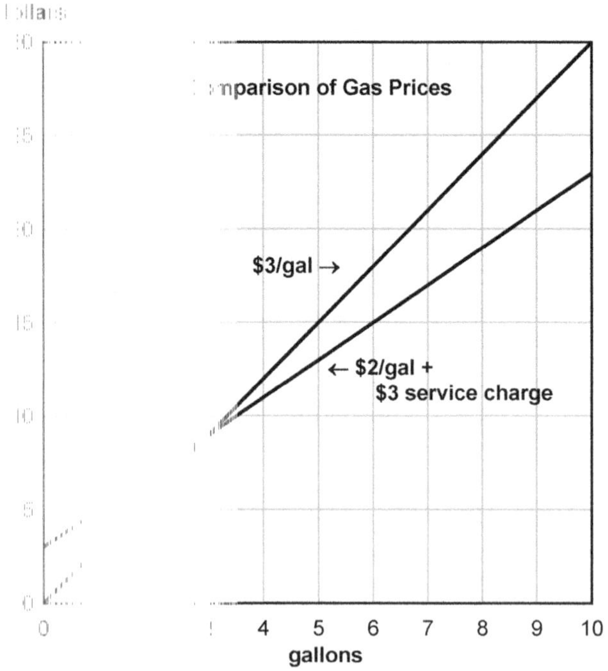

Fig[illegible]uations

The slope of the line, m, is determined by $(y_2 - y_1)/(x_2 - x_1)$. For the two lines representing gasoline prices in Fig. 4-1, we have

1) $$gas\ cost = \frac{\$2}{gal}(x\ gal) + \$3\ pump\ charge$$

$$slope = \frac{(\$9 - \$3)}{(3\ gal - 0\ gal)} = \frac{\$6}{3\ gal} = \frac{\$2}{gal}$$

2) $$gas\ cost = \frac{\$3}{gal}(x\ gal)$$

$$slope = \frac{(\$9 - 0)}{(3\ gal - 0)} = \frac{\$9}{3\ gal} = \frac{\$3}{gal}$$

Note that we have chosen the x axis interval to be from 0 to 3 gallons. This choice makes for an easy calculation of the $\Delta y/\Delta x$ slope. But *any* interval along the plotted line (with its Δy and Δx values) would give the correct slope, because the slope is *constant* at all points on a linear plot.

The linear equation is also sometimes seen in another form, $ax + by = c$. This is often called the **"intercept"**equation. Let the equation be $6x + 2y = 4$. Then, for $x = 0$, the *y intercept* is 2, and the coordinates of the point (which is *on* the line) are $x = 0, y = 2$. For $y = 0$, the x intercept is $4/6$, and the coordinates of this point are $x = 0.66..., y = 0$. Drawing a straight line through these two intercept points then generates the line, $6x + 2y = 4$.

Linear equations are often spoken of as being **simultaneous equations**. This has nothing to do with *time*. Rather, it is because the lines generated by the equations intersect at one certain point. Any two straight lines (which are not parallel) must ultimately meet at an intersection. And when the lines intersect, the coordinates at the point of intersection give a *simultaneous solution* for the equations of *both* lines. For the "gas" equations in question, that intersection occurs at $x = 3\ gal, y = \$9$.

A curve and straight line, or two curves, may also (but do not have to) intersect at one or more places. Consider the equations, $y = x^2$, $y = x + 2$, and $y = x - 2$, plotted in Fig. 4-2. In Fig. 4-2, the line $y = x + 2$ intersects the curve $y = x^2$ at the two points, $x = -1, y = 1$ and $x = 2, y = 4$. These coordinates are the simultaneous solution for the two equations in question. But the line generated by $y = x - 2$ does not intersect the curve, so *no* simultaneous solution exists for the equations $y = x^2$ and $y = x - 2$.

We can confirm the two simultaneous solutions by algebra. If $y = x^2$, and $y = x + 2$,

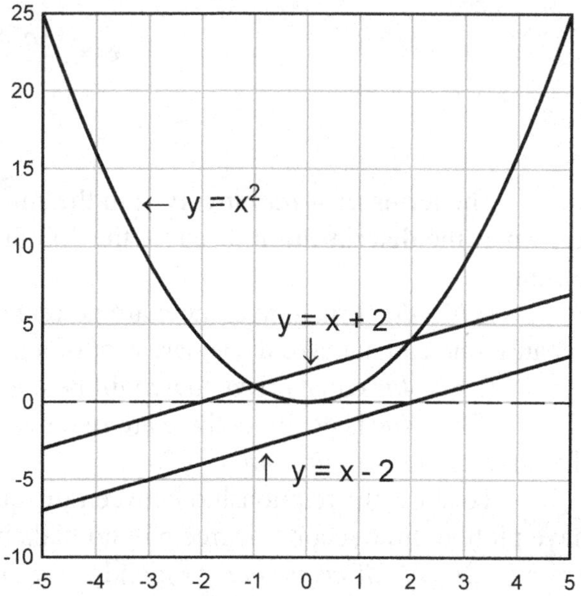

Fig. 4-2 Straight Line-Curve Intersection

then $x^2 = x + 2$, or collecting terms on the left, $x^2 - x - 2 = 0$. This is a *second-degree polynomial* which we can solve for x by the *quadratic formula* (see below) which gives the solutions for x to be -1 and 2. Introducing these values of x back into either equation shows that when $x = -1$, y is equal to 1, and when $x = 2$, y equals 4. These are the coordinates of the intersections seen in Fig. 4-2.

Note that setting the equation for the non-intersecting line, $y = x - 2$, equal to $y = x^2$ yields the second-degree polynomial, $x^2 - x + 2 = 0$. When analyzed by the quadratic formula, this polynomial yields a *negative* determinant (see below) which indicates that no real solution (a solution expressible in *real* numbers) for the polynomial exists. This agrees with the graphic results in Fig. 4-2.

Polynomial Equations

The term, **polynomial**, implies that the equation consists of "several terms." But the definition is actually more restrictive. The mathematical definition of a polynomial is " ...a mathematical expression of one or more algebraic terms, each of which consists of a constant multiplied by a variable raised to a *nonnegative integral* power." The operative terms here are "nonnegative" and "integral." The expression, $y = 2x^3 + 5x^2 - 0.67x + 6$, is a third-degree polynomial, for example, but expressions such as, $y = x^{-2} + 6$, and, $y = x^{0.34} + 4$, are *not* polynomials because of the negative, and fractional, powers that they involve.

The first polynomials we will discuss are **second-degree polynomials**. Second-degree polynomial equations are also known as **quadratic equations**. This last name is unfortunate, because quadratic equations do not involve *anything* relating to four. Second-degree polynomials take the form, $ax^2 + bx + c = 0$, which can be solved by the **quadratic formula** (the quadratic formula is derived in the Chap. 5 on algebra),

$$x = \frac{-b \pm \sqrt{b^2 - 4ac}}{2a}$$

The terms $(b^2 - 4ac)$ involved in the square-root section of the quadratic formula are known as the **discriminant**, because they predict the nature of the solution of the quadratic formula:

If $b^2 - 4ac$ is negative, there are no real solutions (i.e., solutions involving *real* numbers) because one can not take the square root of a negative number.

If $b^2 - 4ac$ equals *zero*, there will be one real solution.

If $b^2 - 4ac$ is positive, there are two real solutions, one being $(-b + (b^2 - 4ac)^{0.5})/2a$, and the other being $(-b - (b^2 - 4ac)^{0.5})/2a$.

To show the relationship between the quadratic formula and the plotted equation, we have plotted three second-degree polynomials in Fig. 4-3: $y = x^2 + x + 1$ *(discriminant* $= -3)$; $y = x^2 + 2x + 1$ *(discriminant* $= 0)$; and $y = x^2 + 3x + 1$ *(discriminant* $= 5)$.

Notice that the curve, $y = x^2 + x + 1$, never has a value of $y = 0$, and thus never intersects the *x axis*. We can solve the equation $x^2 + x + 1 = 0$ by the quadratic formula, but both solutions are *imaginary*, since they involve $\sqrt{-1}$, which is the *imaginary number, i.* As we see,

$$x = \frac{-1 \pm \sqrt{1^2 - 4 \cdot 1 \cdot 1}}{2 \cdot 1} = \frac{-1 \pm \sqrt{-}}{2}$$

$$solns = \frac{-1 + \sqrt{3} \cdot i}{2} \quad and \quad \frac{-1 - \sqrt{3} \cdot i}{2}$$

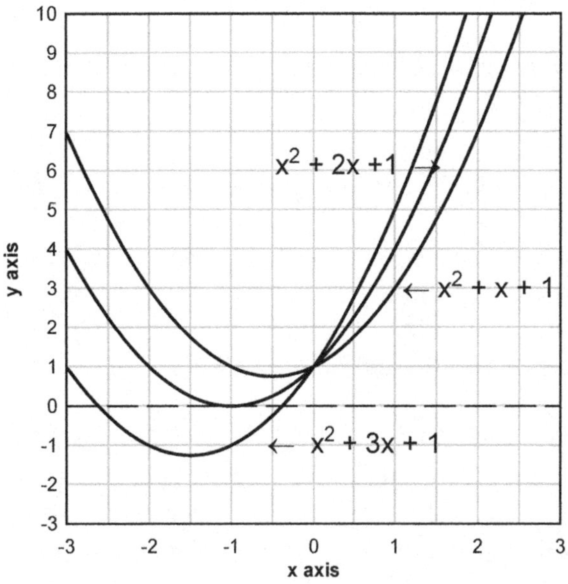

Fig. 4-3 Second-Degree polynomial curves

In the second curve, $y = x^2 + 2x + 1$, the discriminant is equal to *zero*, and we have,

$$x = \frac{-b \pm \sqrt{0}}{2a} \quad and\ (single)\ soln = \frac{-2}{2} = -1$$

and the solution here is *-1*, and it is a *real* number, but there is only *one* solution. (Some authors consider that there are two solutions, but they are *equal*, when the discriminant = *0.)* And in Fig. 4-3, the plotted curve for $x^2 + 2x + 1$ does just touch the *x* axis at *-1* for $y = 0$.

In the case of the third equation, $y = x^2 + 3x + 1$, the discriminant is equal to *five*, and the solutions by the quadratic formula are

$$x = \frac{-3 \pm \sqrt{5}}{2}$$

$$solns\ are \quad \frac{-3 + 2.236}{2} = -0.382 \quad and \quad \frac{-3 - 2.236}{2} = -2.618$$

and we see in Fig. 4-3 that the plotted curve for $x^2 + 3x + 1$ does appear to intercept the *x axis* at these values of *x*.

Second-degree polynomial equations can therefore be solved either by using the quadratic formula *or* by plotting the equation. A second-degree polynomial may have either *zero, one or two* "real"solutions, i.e., solutions that involve only real numbers. Equations that have no real solution will have a negative discriminant, and the plotted equation will not intercept the *x axis* ... on ha... sol... criminant is equal to *zero*, and the plotted equation ... ch the ... quadr... o real solutions show a discriminant greater ... d inter... *axis* ... ints, which are the two solutions of the equation.

Third-Deg... her Po...

The solving ... gree and ... olyn... lgebra is difficult. The Internet provides a number ... h exp... hich ... polynomials can be solved, but it is not simple. The ... sites th... mpu... s which solve third- and fourth-degree polynomials ... lly.

An approac... er to ... olvi... and the use of **factors**. The factors of an expres... terms ... mult... roduce that expression. For instance, *6* has the ... *3*, and ... lying ... or *1 x 2 x 3* yields *6*.

Algebraic e... tor in ... ay, ... me advantages. We know, for example, that the e... $6x^2$ ... *ee* factors involving *x*, since x^3 can only be produc... lying ... e., ... And if the equation is equal to *0*, then the product ... factors ... equ... rite this as

$(x$ ... $)(x$ ...

... $)$ *is* ... *tor* ... *the* n^{th} ***root***

An **important poi**... t if *(x* ... *(x* ... at least one of the factors *must* be equal to *zero*. Th... great... ing ... s.

Let us look ... to ... now the answer. If we ... *2* and ... *3*, the factors multiply ...

$(x-1)(x$ ... $3x+2$

$(x-3)(x^2-3x$ ... x^2+1 ...

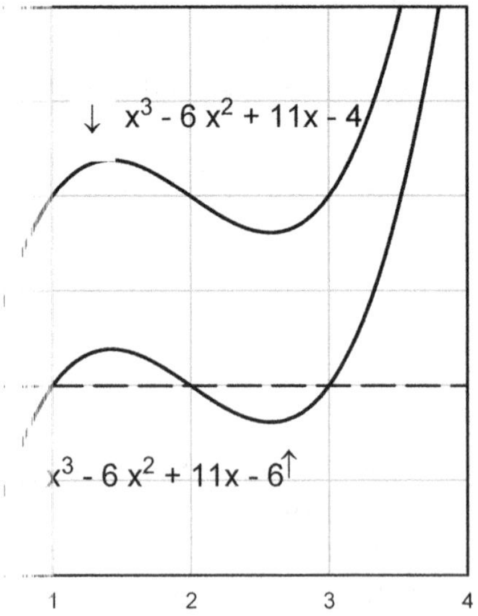

...ntercepts of third-degree curves

We have pl... e, *y* ... *11x - 6*, in Fig. 4-4. ... cepts ... are *x = 1*, *x = 2*, and ... tituti... es into the original equ... an ... *each* case. The *x* in... e plo... are thus the *roots* of th... d rep... three real solutions. This ... d sin... produced the equat... 2 and ... constants in the fac... $- r_2)$ ...

There is a second curve plotted in Fig. 4-4 which represents $y = x^3 - 6x^2 + 11x - 4$. We have just added *2* to the first equation, which changes the constant from *-6* to *-4*. This raises the second curve by *2*, and causes it to have only *one* intercept on the *x* axis at ~0.479 (estimated originally from the graph, and refined by trial-and-error calculation of $x^3 - 6x^2 + 11x - 4 = 0$).

We thus have only one real solution at the *x* intercept of *~0.479*. But this solution represents one factor, i.e., *(x - 0.479)*, of the equation. And we can divide the equation by this factor to obtain a second-degree polynomial that is the *product* of the two remaining factors,

$$\frac{x^3 - 6x^2 + 11x - 4}{(x - 0.479)} = x^2 - 5.521x + 8.355$$

This second-degree polynomial, $x^2 - 5.521x + 8.355$, can then be solved by the quadratic formula,

$$x = \frac{-5.521 \pm \sqrt{(-5.521)^2 - (4 \cdot 1 \cdot 8.355)}}{2 \cdot 1} = \frac{-5.521 \pm \sqrt{-2.939}}{2}$$

$$x = -2.761 + 0.429\,i \quad \textit{and also} \quad x = -2.761 - 0.429\,i$$

The equation has three solutions also, but only one is real. The other two involve the imaginary number $i\ (= \sqrt{-1})$, and are therefore *imaginary*. The real solution is the irrational number *0.479*, but it satisfies the equation in question

$$\begin{array}{lllll} x^3 & -\ 6x^2 & +\ 11x & -\ 4 & = 0 \\ (0.479)^3 & -\ 6(0.479)^2 & +\ 11(0.479) & -\ 4 & = ? \\ 0.1099 & -\ 1.3766 & +\ 5.269 & -\ 4 & = -0.002.. \approx 0 \end{array}$$

And we could make *0.002..* even closer to *0* by taking *0.479* to more decimal places.

Fourth-degree and higher polynomials are more difficult, but can be treated in the same manner. By plotting the equation, we can see if the resultant curve intercepts the *x axis* at any point. If so, we can estimate the value of the real number solution(s) from the intercept points. We can then use these solutions as *factors* to divide the original equation, and possibly obtain a second-degree polynomial (as above). Any second-degree polynomials that are obtained can then be solved by the quadratic formula.

A polynomial *may* have *up to* as many solutions as its degree. For example, fourth-degree polynomials can have as many as four solutions, but they may also have only *zero, one, two* or *three* solutions. If the curve produced by plotting a polynomial does not intercept the *x axis*, the polynomial has *no* real solutions.

Equations with x in the Denominator

When x appears in the denominator of an equation, there is always the possibility for "division by zero", which is *not* allowed. The simplest equation with x in the denominator is

$$y = \frac{1}{x}$$

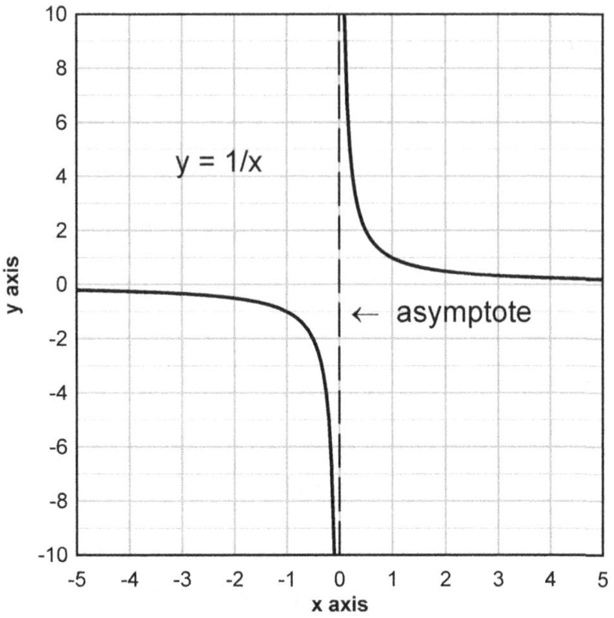

Fig. 4-5 Curve of y = 1/x

If we plot this function (Fig. 4-5), we see that the curve is *not* continuous, and that a break in the curve occurs at $x = 0$. As x comes close to 0 from the negative direction, y becomes more and more increasingly negative. As x approaches 0 from the positive direction, y becomes an increasingly large positive number. This results from the reciprocal nature of the function.

Note that each side of the discontinuous curve approaches, but never touches, a limiting line at $x = 0$, which is known as the **asymptote** (from a *Greek* word meaning "not meeting"). No matter how large x may become, y never becomes equal to 0, but always has some discrete value even if exceedingly small.

This same general behavior occurs for all divisions by x (or expressions containing x). As an example, consider

$$y = \frac{x^3 - 6x^2 + 11x - 6}{x - 4}$$

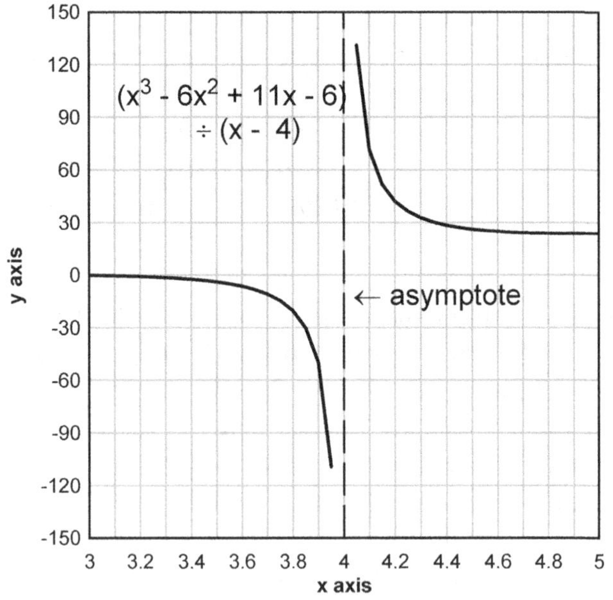

Fig. 4-6 Division of Polynomial by (x - 4)

When this function is plotted (Fig. 4-6), we see much the same behavior as in Fig. 4-5. But now the discontinuity in the curve falls at $x = 4$, because this is the point at which $x - 4$ is equal to 0.

In other cases, the dividend and divisor may be more complex than above, and include higher degree polynomials. This additional complexity may add peaks to the curve and change its overall form. However, it will *always* be the case that when *x* in the divisor (or an expression containing *x*) becomes equal to *0*, the curve will tend toward negative or positive infinity, and approach an asymptote just at the point where $x = 0$.

Addition of sin Curves

Certain trigonometric functions such as *sin (x)* produce *repetitive* curves with values that vary between *1* and *-1*. If we plot *sin (x)* and *sin (2x)*, we see that the *sin (2x)* produces twice the number of peaks as *sin (x)*. That is, the *sin (2x)* peaks occur at twice the *frequency* of the *sin (x)* peaks. If we plotted *sin (10x)*, we would find *10* times as many peaks in the same *x* interval.

Moreover, we can add the *sin (x)* and *sin (2x)* curves and obtain a composite curve as shown in Fig. 4-7. And, by adding *sin* curves of different frequencies, it is possible to obtain wave forms that reproduce the sound of musical notes. This is the principle behind musical synthesizers. In mathematics, a function that produces a repetitive wave form is known as a **periodic function**, and under certain conditions, any periodic function can be represented by a **Fourier series**.

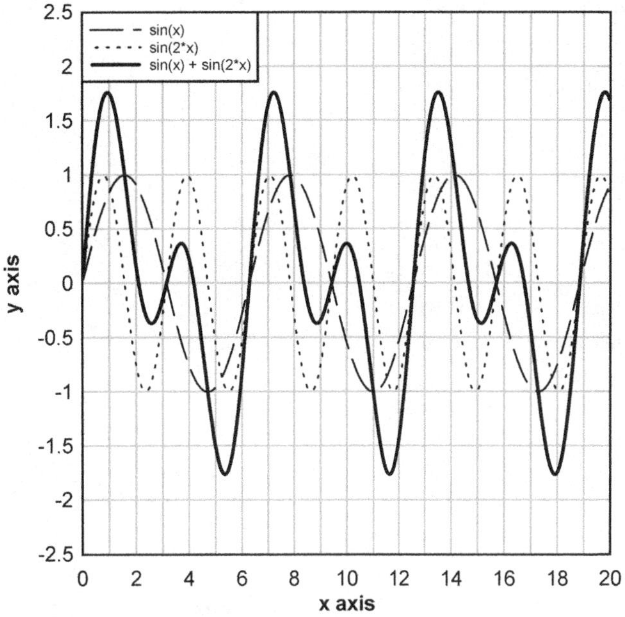

Fig. 4-7 Addition of sin curves

Closing note

In the first four chapters of this book, we have looked at the simple concepts and calculations that provide the basis for lower-level mathematics. In the chapters to come, we will apply the ideas we have developed so far to various areas of mathematics. There is a considerable redundancy in this approach. We "talk about something" and then we "talk about it again." But I believe the repetition is justified. Very few of us can be told something just once, and truly grasp and understand it.

Chapter 5.

Basic Algebra

In algebra w [illegible] numb [illegible] va [illegible] *ot* know. However we can use **equations** (i.e., con [illegible] quality [illegible] hese [illegible] (or **variable)** numbers to known numbers. Take the [illegible] un I?" [illegible] We [illegible] d, for example, that if we multiply Jim's age [illegible] btract [illegible] r fo [illegible] Bill's age, the answer is *6*. And also that the sum of [illegible] l's ag [illegible] ting [illegible] e *x*, and Jim's age be *y*, we can set this information [illegible] of alg [illegible]

3 [illegible] 6

and [illegible] 7

These two e [illegible] n be sol [illegible] *po* [illegible] qual sign. If we multiply both sides of an equation [illegible] *e* num [illegible] s [illegible] qual. Moreover, since the two sides are equal, we [illegible] equal [illegible] *ond* [illegible] nd obtain a third equation whose two sides are still e [illegible]

x eq [illegible] = 14

add [illegible] = 6

di [illegible]

[illegible] *eqn* [illegible] 4) [illegible] *or* $y = 3$

Therefore Bill (x) is [illegible] , and Ji [illegible] is [illegible] year

Algebra is [illegible] ess. [illegible] ents [illegible] each chess piece, and the general rules of che [illegible] le, an [illegible] plain [illegible] minutes. But chess masters devote their entire li [illegible] tudy [illegible] of [illegible] chess move is simple, but obtaining the corre [illegible] *on* of [illegible]

In algebra, [illegible] tical [illegible] nvol [illegible] very limited in number. We have only the prope [illegible] *ity*; [illegible] ic st [illegible] tion, subtraction, multiplication and division; and n [illegible] divis [illegible] oner [illegible] ns. Mathematicians over the centuries, however, [illegible] ed as [illegible] sult [illegible] ra using only these basic tools.

We will now [illegible] re of [illegible] pro [illegible] are encountered in beginning algebra.

One unkno [illegible] uation

To solve for [illegible] n, *x*, w [illegible] e a [illegible] uation that relates *x* to known numerical values. [illegible] quation [illegible] can [illegible] e value of *x*. Equations of one unknown can be br [illegible] rthe [illegible] *egre* [illegible] nown.

First degree. Equations where the highest exponent of x is x^1 are known as "first degree" equations. Many first degree equations are simply "arithmetic" equations such as

$$x - 4 = 6 \qquad 2x - 3 = 11 \qquad 2(x + 2) = 8 \qquad \frac{(x - 4)}{2} = 4$$

Equations of this type may be more complex than shown here, and involve larger and not so simple numbers, but they can all be solved by just the four arithmetic operations (which must, of course, be applied to *both* sides of the equation).

The unknown, x to the first degree, can also be involved in exponential calculations. Equations where x is in the exponent, or in the root symbol, often "look funny" to students . But if we "raise to the power" or "take the log (or ln)" of both sides, using our calculator, they can all be solved. For example,

$$e^x = 6, \textit{ then } \ln e^x\ (= x) = \ln 6, \textit{ or } x = 1.7917...$$

$$10^{2x} = 48, \textit{ then } \log 10^{2x}\ (= 2x) = \log 48, \textit{ or } 2x = 1.6812..., x = 0.8406...$$

$$\frac{1}{e^{2x}} = 4, \textit{ then } \ln e^{2x}\ (= 2x) = \ln \frac{1}{4}, \textit{ or } 2x = -1.3862..., x = -0.6931...$$

$$\sqrt[x]{32} = 2, \textit{ then } (\sqrt[x]{32})^x = 2^x, \textit{ or } 32 = 2^x,$$

$$\textit{then } \ln 32 = x \ln 2, \textit{ and } x = \frac{3.4657...}{0.6931...} = 5$$

In the last example, student might not recall why $\ln 2^x = x \ln 2$. This comes from the definition of multiplication by exponents, i.e.,

$$x^3 = x \cdot x \cdot x = \ln x + \ln x + \ln x = 3 \ln x$$

Second degree. In second degree equations, the highest exponent of x is *2* (i.e., x^2). Second degree equations may also contain x^1 terms. Second degree equations are commonly referred to as "quadratic" equations, although (except in historical terms) they have nothing to do with the number *4*. Another name is a "second-degree" polynomial (see **Polynomial Equations**, pg. 44) The general form of a second-degree polynomial is

$$ax^2 + bx + c = 0$$

Note that many quadratic equations may *initially* appear to be more complex than this. But if the highest degree of x in the equation is x^2, it can always be reduced to the quadratic form, although the coefficient, b, and the constant, c, may be *zero* in some cases.

Development of quadratic formula. The **quadratic formula** allows you to solve *any* second-degree polynomial equation. It is a very powerful tool in mathematics. One derivation of the quadratic formula is based on a technique for solving equations known as **"...completing the square"**, which we show here.

We begin by rewriting the quadratic equation as follows,

$$ax^2 + bx + c = 0$$
$$\downarrow$$
$$x^2 + \frac{b}{a}x + \frac{c}{a} = 0$$
$$\downarrow$$
$$x^2 + \frac{b}{a}x = -\frac{c}{a}$$

Next we look at the standard form for the quadratic equation, i.e.,

$$(x + h)^2 = (x + h)(x + h) = x^2 + 2hx + h^2$$

Comparing these two equations, we see that the coefficient for x is b/a in the first equation, and $2h$ in the last equation. Therefore, $2h = b/a$, or $h = b/(2a)$. And then $h^2 = (b/2a)^2 = b^2/4a^2$. Adding "$h^2$" to the left side of the original equation will "complete the square" on that side. But to keep the equation *equal*, we must also add $b^2/4a^2$ to the right side of the equation. This gives

$$x^2 + \frac{bx}{a} + \frac{b^2}{4a^2} = \frac{-c}{a} + \frac{b^2}{4a^2}$$

The left side of the equation is thus equal to the *square* of $(x + b/2a)$. And the *common denominator* of the right side is $4a^2$, so we have

$$\left(x + \frac{b}{2a}\right)^2 = \frac{-4ac + b^2}{4a^2} = \frac{b^2 - 4ac}{4a^2}$$

Taking the square root of both sides, and reducing the resultant equation then yields the quadratic formula,

$$x + \frac{b}{2a} = \pm \sqrt{\frac{b^2 - 4ac}{4a^2}}$$

or

$$x = \frac{-b \pm \sqrt{b^2 - 4ac}}{2a}$$

As discussed in the plotting of **polynomial equations** (pg. 45), the expression $(b^2 - 4ac)$ in the quadratic formula is the **discriminant**, and it predicts the nature of the solution(s) of the quadratic formula:

If $b^2 - 4ac$ is negative, there are no real solutions;
If $b^2 - 4ac$ equals *zero*, there is one real solution;
If $b^2 - 4ac$ is positive, there are two real solutions.

Third degree and higher polynomials. As mentioned before (pg. 46), there is no *simple* equation equivalent to the quadratic formula for solving third degree and above polynomials.

As shown in the previous Chapter for third-degree polynomials, one of the primary means for solving high-degree polynomials is by **factors**. If we have a polynomial such as

$$x^4 + Ax^3 + Bx^2 + Cx^1 + D = 0$$

we know that this polynomial can be viewed as being the product of *four* factors containing x, i.e.,

$$x^4 + Ax^3 + Bx^2 + Cx^1 + D = (x - r_1)(x - r_2)(x - r_3)(x - r_4) = 0$$

This is because the term x^4 *must* be produced by the multiplication of four x terms, or $x \cdot x \cdot x \cdot x$. Moreover, if the product of a series of $(x - r_n)$ factors is equal to 0, then *at least* one of the factors must itself be equal to 0. If we have that

$$(x - r_1) = 0, \ \textit{then} \ x = r_1$$

The factors can be difficult to determine. One of the best approaches is to plot the polynomial. This can be easily done with present day high-speed computers and inexpensive plotting software (this book uses *DPlot* which cost about $50). Let us consider two curves in Fig. 5-1 which are plotted from equations that we have composed, and so know their solutions:

Curve 1 [illegible] $2)($ [illegible] $1)$ [illegible] $+ 2x^2 - 5x - 2 = 0$

Curve 2 [illegible] $2)($ [illegible] $1) =$ [illegible] $3x - 2 = 0$

Curve 1 intercepts [illegible], at [illegible] points on the x axis, [illegible] *-2.6* (estimated), *-2*, *-0.* [illegible] and [illegible] The intercepts of *1* [illegible] that [illegible] the factors are $(x - 1)$ [illegible] And [illegible] we divide the origi[illegible] degree polynomial by thes[illegible] we [illegible] the third factor, x^2 [illegible]

This factor [illegible] degree polynomial, which [illegible] by [illegible] quadratic formula,

$$x = \frac{-b \ldots}{\ldots}$$

$$= \frac{-3 \ldots (1)}{\ldots}$$

$$= -0.3\ldots \quad \ldots 518$$

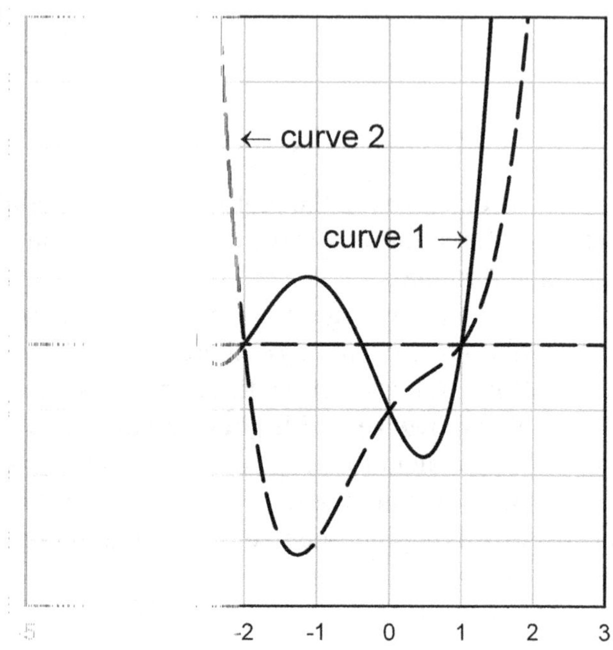

Fig 5-[illegible] fourth-degree polynomials

The discriminant of [illegible] $- 4ac$, [illegible] equa[illegible] is a positive number, and tells us that the solution[illegible] second-degree [illegible] *l* numbers. The solutions to the fourth-degree poly[illegible] general[illegible] are [illegible] all real numbers.

Curve 2 int[illegible] re, y [illegible] points [illegible] axis, these points being *-2* and *1*. We know therefore [illegible] the fact[illegible] four[illegible] polynomial are *(x + 2)* and *(x - 1)*. If we divide these t[illegible] o the [illegible] polyn[illegible] obtain the third factor which is $(x^2 - x - 1)$. This ca[illegible] by th[illegible] form[illegible]

$$x = \frac{-b \pm \sqrt{b \ldots}}{2a} \ldots \frac{1 \pm \ldots}{2} \ldots \frac{1 \pm 3 \ldots}{2} \ldots \text{ are } \frac{1}{2} + \frac{3}{2} i \text{ and } \frac{1}{2} - \frac{3}{2} i$$

There are thus two [illegible] and t[illegible] factors [illegible] curve 2 polynomial.

As a further [illegible] can [illegible] fifth-degree [illegible] nomial by multiplying our two fourth-degree poly[illegible] tions b[illegible] factor. The two fifth-degree polynomials that are result are pl[illegible] 5-2.

The behavior of the curves is just an extension of the plots of the fourth-degree polynomials. For curve 1, there are now *five* intercepts of the *x* axis intercepts, with three being integers at *-2, -1* and *1*, and two being decimals at ~*2.6* and ~*0.4*. The integer values represent the factors *(x +2)*, *(x + 1)* and *(x - 1)*, and the two decimal numbers are the quadratic formula solutions for the residual factor, $x^2 + 3x + 1$. This is just the same as we saw with the fourth-degree curves.

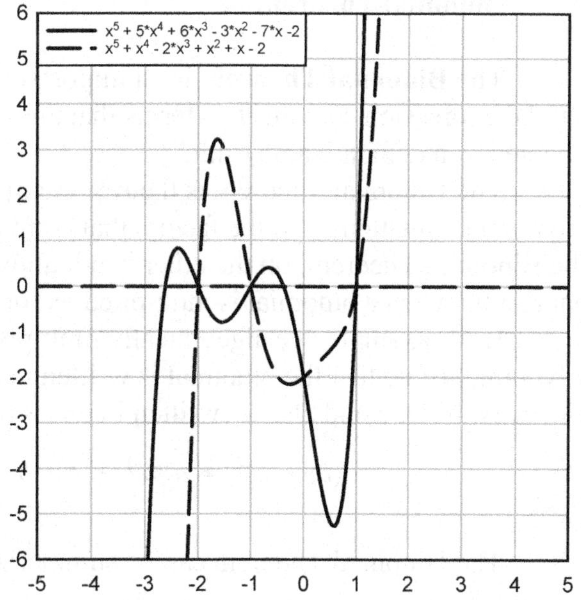

Fig 5-2 Plot of Fifth-Degree Polynomials

For curve 2, there are only *three* intercepts of the *x* axis, these being at *-2, -1* and *1*. These represent the factors of *(x + 2)*, *(x + 1)* and *(x - 1)*. And if we divide these back into the fifth-degree equation, the remaining factor is $x^2 - x - 1$. And as shown above, the quadratic solutions to this factor are the imaginary numbers,

$$\frac{1}{2} + \frac{1}{3}i, \text{ and } \frac{1}{2} - \frac{1}{3}i$$

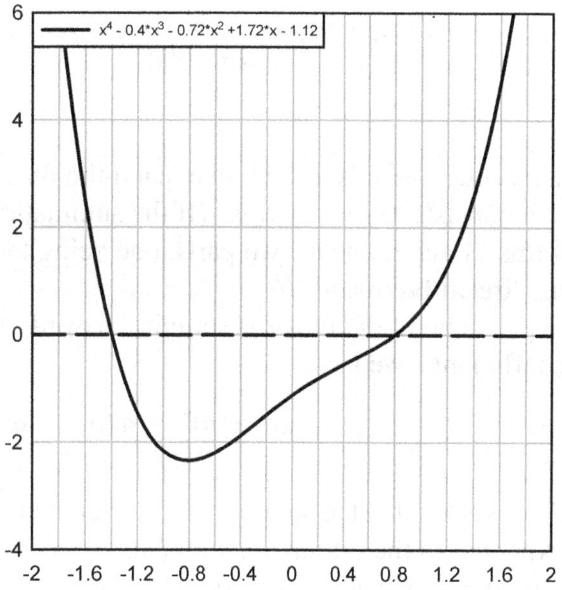

Fig 5-3 Plot of fourth-degree polynomial

One might question if this procedure would work for *non-integer* factors. We can check this by multiplying the quadratic factor of curve 2, $x^2 - x + 1$, by the factors *(x - 0.8)* and *(x + 1.4)*, giving $x^4 - 0.4x^3 - 0.72x^2 + 1.72x - 1.12$, a fourth-degree polynomial plotted in Fig. 5-3. The curve intercepts the *x* axis at *-1.4* and *0.8*, which indicates that the factors are *(x+1.4)* and *(x - 0.8)*.

One may object that these are "made up" examples, but the point that they illustrate is always true for *any* plotted polynomial. The value of the point where the plotted polynomial intersects the *x* axis is a *factor* of the polynomial.

Binomial Theorem

The **Binomial Theorem** is an important part of the history of algebra. "Binomial" means that the expression contains *two* terms that are polynomials, and "theorem" means "... an idea accepted as a demonstrable truth."

The Binomial Theorem is the *reverse* of factoring. In factoring, we resolve complex polynomial equations into the factors that were multiplied to produce the complex equation. The Binomial Theorem, on the other hand, allows us to easily calculate the complex polynomial that results when a binomial is multiplied exponentially, that is, $(a + b)^n = ?$

If for example, we algebraically multiply out the function, $(a + b)^4$, we obtain this polynomial (note that the binomial is written here in terms of *a* and *b*, which would normally be *constants*, but it could also be written in terms of *variables*, e.g., *x* and *y*),

$$(a + b)^4 = a^4b^0 + 4a^3b^1 + 6a^2b^2 + 4a^1b^3 + a^0b^4$$

The Binomial Theorem can be summarized as a mathematical formula,

$$(a + b)^n = \sum_{k=0}^{n} {}^nC_k \, a^{n-k} b^k$$

$$where \quad {}^nC_k = \frac{n!}{(n-k)!k!} = binomial\ coefficient,\ and\ a^{n-k}b^k = literal\ factor$$

This equation tells us that, to expand the function $(a + b)^n$, we must produce a summation series that consists of $n + 1$ terms. (If the summation started with $k = 1$, then there would be only n terms.) Each term has two parts, one being the **binomial coefficient** (given by ${}^nC_k$), and the other the **literal factor**, $a^{n-k} b^k$.

The literal factor is straight forward. If we look again at $(a + b)^4$, and omit the binomial coefficients, we have

$$(a + b)^4 = a^4b^0 + a^3b^1 + a^2b^2 + a^1b^3 + a^0b^4$$

And we see that the sum of the exponents in each term is simply equal to the value of the exponent in the original binomial expression; in this case, *4*. Notice that we obtain *five* terms (i.e., $n + 1$) when $n = 4$, and that the first and fifth terms contain b^0 $(= 1)$, and a^0 $(= 1)$. The first term contains b^0 because we start with $k = 0$, and the fifth term has a^0 because $4 - 4 = 0$. Since they both equal 1, b^o and a^o are usually not shown.

The exponents in the literal factor of any term in a binomial expansion can therefore be calculated easily. But we must be careful that we are speaking of the correct term. Consider that we wish to know the exponents in the literal factor of the 7th term in the expansion of $(a + b)^{10}$. We would have

$$a^{10}b^0 + a^9b^1 + a^8b^2 + a^7b^3 + a^6b^4 + a^5b^5 + a^4b^6 + \cdots$$

The exponent values in the 7th term are therefore not $a^3\,b^7$, but rather "one less" at $a^4\,b^6$.

We now have, if the values of *a* and *b* are known, an easy way to calculate the literal factor in each term of a binomial expansion. But we do *not* have a formula that lets us predict the binomial coefficient. To do this, we must resort to either to combinatorial mathematics, or to Pascal's triangle.

Combinatorial mathematics is a well-developed area of mathematics, but it is probably best known to the public as a means of calculating the odds of winning a lottery. Many lotteries involve picking *6* winners out of *49* total numbers. The player first picks *1* number out of *49*, then *1* out of the remaining *48*, *1* out of the remaining *47*, and so on. The odds of finding a winning number in the first pick are *1* in *49* (or *49/1*), for the second pick, *1* in *48* (*48/1*), for the third pick, *1* in *47* (*47/1*), and so on. The odds of any two events happening is simply the odds for the first event *times* the odds for the second event, so the *preliminary* odds for picking six winning numbers are

$$(49 \cdot 48 \cdot 47 \cdot 46 \cdot 45 \cdot 44) \textit{ to one} = \textit{preliminary odds of winning}$$

However, the picks do not have to be in order. That is, the first pick does not have to be, say, the lowest number of the six winners. The ways in which a given number of items can be combined turns out to be the **factorial** of that number. For example, suppose that we need only three winning numbers, and they are *1*, *2* and *3*. We could pick these numbers as *123*, but also as *132*, *213*, *231*, *312*, and *321*. For three numbers, there are six ways in which they can be combined, and *6* is the factorial of *3!* (i.e., $1 \cdot 2 \cdot 3 = 6$). When *six* numbers are to be picked, they can be chosen in *6!* ways, so the odds for picking *6* numbers out of *49* is made *better* by a factor of *6!*, giving

$$\frac{(49 \cdot 48 \cdot 47 \cdot 46 \cdot 45 \cdot 44)}{6!} = \textit{odds of winning}$$

We could simply multiply this out on our calculator. However, the combinatorial number is usually presented in general form. The product of $49 \cdot 48 \cdot 47 \cdot 46 \cdot 45 \cdot 44$ is equal to *49!* divided by *43!* But *43!* is equal to the factorial of *(49 - 6)* where *49(or n)* is the total numbers and *6 (or k)* is the number of possible ways of picking the winning *6* numbers. We can therefore write

$^{49}C_6$ = [illegible] 6!(4[illegible] [illegible] 13,[illegible] [illegible]*dds against winning*)

[illegible] *general* [illegible] [illegible] *k*)!

The combinatorial [illegible] s us[illegible] [illegible]bilities), chose *k* ", and it can be calculated by m[illegible] calcu[illegible]

We have in[illegible] **fa**[illegible] [illegible]bers are defined as the product of the multiplicatio[illegible] *posi*[illegible] [illegible]nd member, *n*. Factorial numbers are writte[illegible] sider [illegible]rial,

7 *fac*[illegible] ! = 1·2[illegible] = 50[illegible] [illegible] = 4·1·6·3·7·5·2

The numbers in the[illegible] throu[illegible] *fac*[illegible] [illegible]nal product number, *5040.* Notice that the fact[illegible] ve to b[illegible] [illegible]gh *7* in sequence. Any order of multiplication will [illegible] same [illegible]ctori[illegible] [illegible]wn in the *1 through n* sequence for clarity, and con[illegible] calcul[illegible] other [illegible]umbers. Before scientific calculators, factori[illegible] in t[illegible] mathem[illegible] books. The value of *100!*, for instance, was show[illegible] *226*[illegible]

Combinat[illegible] and [illegible] Coeff[illegible]

Let us now [illegible] com[illegible] [illegible]nomial coefficients relate. A combinatorial nu[illegible] tes the [illegible] ways [illegible]ick *k* objects from an *n* number of total objects. O[illegible] omia[illegible], and produced the result,

[illegible] a^4b^0 [illegible] + 5[illegible] $b^3 + a^0b^4$

We will let the *n* nu[illegible] jects" [illegible] he w[illegible] [illegible]h we can pick (or arrange) them be *k*, which g[illegible] *4*. ([illegible] can[illegible] [illegible]r than *n.)* This gives

${}^nC_k$	*equals*	$\dfrac{n!}{(n-k)!k!}$	
${}^4C_0$	1	$\dfrac{4!}{(4-0)!0!} = 1$	*n.b.* $0! = 1$
${}^4C_1$	4	$\dfrac{4!}{(4-1)!1!} = 4$	
${}^4C_2$	6	$\dfrac{4!}{(4-2)!2!} = 6$	
${}^4C_3$	4	$\dfrac{4!}{(4-3)!3!} = 4$	
${}^4C_4$	1	$\dfrac{4!}{(4-4)!4!} = 1$	

And we see that the combinatorial values are *just the same* as the binomial coefficients in the binomial expansion equation.

To show how binomial coefficients relate to "making choices", let us consider the following expansion for $(x + y)^4$, where the x and y subscripts indicate *which* factor the term comes from,

$$let\ (x + y)^4 = (x_1 + y_1)(x_2 + y_2)(x_3 + y_3)(x_4 + y_4)$$

If we multiply out the factors, we will find 2^4 (i.e., *16*) terms. There will be *only 1* term that contains nothing but x's. That will be $x^4 = x_1 x_2 x_3 x_4$, and it is formed by combining the x from *each* of the four factors. Its combinatorial number is given by ${}^4C_0 = 1$.

There will be *4* terms of the "form" of x^3y, these being

$$x_2x_3x_4(y_1) \quad x_4x_3(y_2)x_1 \quad x_4(y_3)x_1x_2 \quad (y_4)x_1x_2x_3$$

The number of terms of "x^3y" is therefore seen to be equal the number of ways of choosing one component, y, from the *4* components of the term. Its combinatorial number is ${}^4C_1 = 4$.

There will also be *6* terms of the "x^2y^2 "form. These will be

$$(y_1y_2)x_3x_4 \quad (y_2y_3)x_1x_4 \quad (y_1y_3)x_2x_4 \quad (y_2y_4)x_1x_3 \quad (y_1y_4)x_2x_3 \quad (y_3y_4)x_1x_2$$

This represents all of the ways of taking *two* things from *four*. The combinatorial number is thus ${}^4C_2$ which is equal to *6*.

There will then be 4 terms of the "xy^3 " form. These will be just the same as the "x^3y" form, but with x and y interchanged. Their combinatorial number will be ${}^4C_3$ which equals 4.

Finally, there will be 1 term of y^4 or $y_1\ y_2\ y_3\ y_4$. The total number of all of the terms is $1 + 4 + 6 + 4 + 1 = 16$, which is equal to the number of terms predicted by $2^4 = 16$.

Therefore while we only think of "multiplying out" the 4 initial factors,

$$(x_1 + y_1)(x_2 + y_2)(x_3 + y_3)(x_4 + y_4)$$

the process is equivalent to choosing k numbers from n total numbers, and so the coefficients that result can be derived from combinatorial numbers.

Pascal's triangle and probability

Although partially known in ancient times, **Pascal's triangle** is considered in the West to have been developed by **B. Pascal** (*1623-1662*) in about *1650*. Over the years, Pascal's triangle has been extensively studied, and many unique patterns in its numbers have been identified. We are interested here, however, only in how Pascal's triangle provides the binomial coefficients, and how it predicts the probability of random events occurring.

Here is an abbreviated version of Pascal's triangle. Each number is the sum of the numbers to its *upper left* and *upper right*. In Row 4, for example, the number *4* is the sum of *1* plus *3*, the number *6* is the sum of *3* plus *3*, and the second number *4* is the sum of *3* plus *1*. The same scheme holds for Row 5, and beyond.

1 Row 0
1 1 Row 1
1 2 1 Row 2
1 3 3 1 Row 3
1 4 6 4 1 Row 4
1 5 10 10 5 1 Row 5

Pascal's Triangle

To produce the end members of *1* in each Row, we consider that there is a *zero* (which is not shown) at each end of each Row, including Row 0. The summation is then $0 + 1 = 1$, and $1 + 0 = 1$ for the visible ends of the Row which lies below.

Inspection shows that the numbers in the different rows of Pascal's triangle are the binomial coefficients for a binomial raised to the power that is the number of the Row. Row 4 corresponds to $(x + y)^4$, and numbers of *1, 4, 6, 4, 1* are indeed the binomial coefficients of the expansion of $(x + y)^4$. The same is true for Row 2 and Row 3, and Row 5 and higher as well (if one multiplies out the binomial expansion).

To explain this, remember that each binomial coefficient corresponds to how many ways a certain number of objects can be chosen from a larger set of objects, i.e., the combinatorial number, ${}^nC_k$. It is, in another sense, simply a matter of probability.

The role of Pascal's triangle in predicting probability can be demonstrated by the flipping of a coin. Let the coin be flipped 4 times. There are $4^2 = 16$ sequences of results that could occur. Letting H be heads, and T be tails, these are

(H,H,H,H)	1
(H,H,H,T),(H,H,T,H),(H,T,H,H),(T,H,H,H)	4
(H,H,T,T,),(H,T,T,H),(T,T,H,H),(H,T,H,T),(T,H,T,H),(T,H,H,T)	6
(H,T,T,T),(T,H,T,T),(T,T,H,T),(T,T,T,H)	4
and (T,T,T,T)	1

These are the only possibilities, and they sum up to *16*. And they indicate these probabilities. The odds of flipping any of the *exact* sequences, including (H,H,H,H) and (T,T,T,T), is *1 in 16*. Disregarding sequence, the odds of flipping *3* heads and *1* tail (or *3* tails and *1* head) is *4 in 16*, and the odds of flipping *2* heads and *2* tails is *6 in 16*.

The binomial theorem and Pascal's triangle are an important part of mathematical history. But lower-level algebra is mostly concerned with solving algebraic equations to find applied answers. At the beginning of this Chapter, we examined some means for solving algebraic equations having *one* unknown. We will continue on now to look at equations involving *two or more* unknowns.

Two unknowns, two equations...

You can not solve for two unknowns from a single equation. Consider $x + y = 4$. If $x = 2$, then $y = 2$; if $x = 1, y = 3$; if $x = 0.5, y = 3.5$, and so. There is an *infinite* number of solutions, but no single *unique* solution.

A second *independent* equation that relates x and y is required. For example, if we have a second equation such as, $x - y = 1$, we can combine it with the original equation, and find the solution, e.g.,

$$\begin{array}{rl} x + y = 4 & \textit{eqn } (1) \\ + \; (x - y = 1) & \textit{eqn } (2) \\ \hline 2x = 5, \quad \textit{or } x = 2.5 & \\ \textit{and then} \quad y = 1.5 & \end{array}$$

In the second equation, note that y is *always* equal to x-*1*. We can thus **eliminate** y by adding eqn (2) to eqn (1), because we are adding *equal* value to each side of eqn (1). Subtraction of eqn (2) from eqn (1) would also work, but in that case we would be determining y first.

Alternatively, we could **substitute** the x value for y *(i.e., x-1)* in place of every y in eqn (1), and then solve eqn (1) for the single unknown, x. **Elimination** or **substitution** are the main approaches in algebra to solving equations.

If a means f... uation... available, systems ... two eq... can often be solved ... The ... equations above ha... ted in ... They are first-degr... nom... generate straight li... tersec... the point of interse... occur... 2.5 and $y = 1.5$, is ... ous s... both equations. Re... ight li... intersect unless the... l or o... lapping. Also note ... tion i... same as we obtaine... alge...

Nonlinear e... fferen... can also sometimes ... graph... their plotted curves ... Fig. ... have plotted two c... ne cur... determined by a se... polyn...

$y = x^2$ ... 2

and the second cur... by th... logarithmic functi...

$y = $ ...

The curves intersec... rs, on... $x = 2.9$ and $y = 5.0$... curs... original graph), an... appr... $x = 0.0025$ and $y =$ ...

We can veri... vers ... substituting them i... al eq...

if y ... + ...

then y ... $= 0$

and 5.03 − ln(... = ~0...

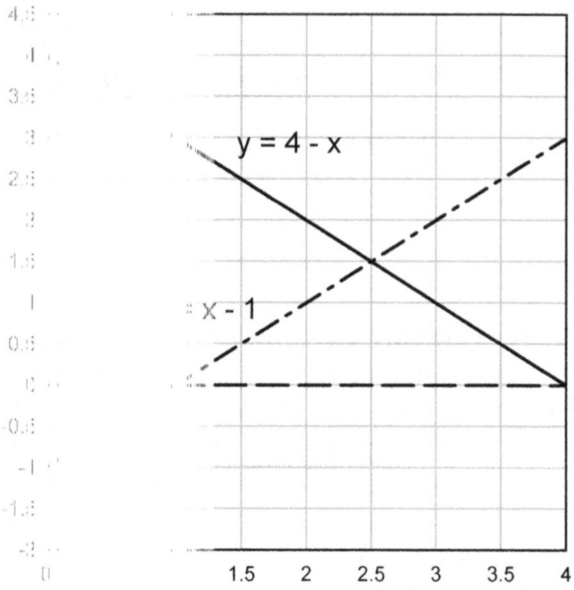

Fig ... laneous solution of two lines

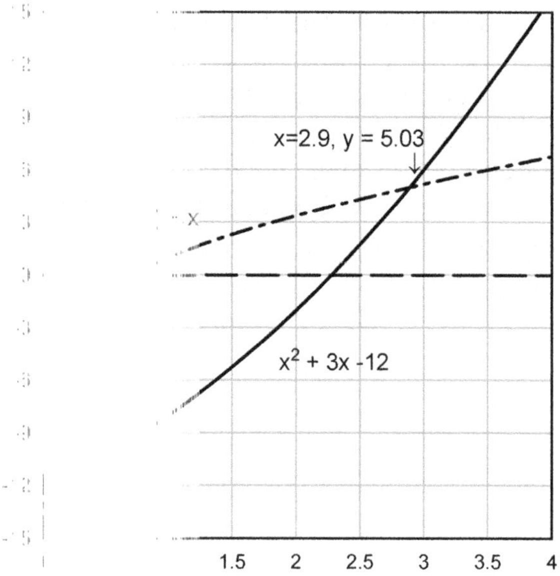

F... of dissimilar curves

The "residual error" of ~ 0.001 is close to 0, and the difference can be made even smaller by further reiterative calculations.

For the second solution where $x = 0.0025$ and $y = -11.98$, we will test the polynomial equation

$$y = x^2 + 3x - 12$$

$$\textit{with } y - x^2 - 3x + 12 = 0$$

$$\textit{and } -11.98 - (0.0025)^2 - 3(0.0025) + 12 = \sim 0.012$$

and find a residual error of ~ 0.012. If the y value was -11.99, the residual error would be lowered to ~ 0.0025. The algebraic solutions thus agree with the graphic solutions.

We can also reduce the equations to the "one equation, one unknown" form by substituting for y, i.e.,

$$x^2 + 3x - 12 = \ln(x^2) + x,$$
$$\textit{which reduces to: } x^2 + 2x - \ln(x^2) = 12$$

But, interestingly, this equation, $x^2 + 2x - \ln(x^2) = 12$, can *not* be solved by simple algebra as performed by several algebraic equation solvers offered on the Internet. (See, for example, *www.algebrahelp.com/calculators*.) Mathematicians recognize that many valid algebraic equations can not be solved analytically. However such equations can often evaluated by graphing, as here, or by trial-and-error calculation.

Three unknowns, three equations...

Systems of involving three unknowns and three equations are usually presented using first-degree variables, which are also often described as *linear* variables. This can be somewhat misleading because, if you plot an equation such as

$$x + y + z = C$$

the resultant figure is not a line, but rather a *plane;* that is, a figure in three dimensions. The "simultaneous solution" represented by the intersection of the two planes produced from the equations,

$$x + y + z = C$$

$$x + 2y + z = D$$

is a *line* in three-dimensional space, and not a single point. A single-point solution, having the unique coordinates of x, y and z, is found only at the point of intersection of *three* planes.

Because of the difficulty of plotting three-dimensional graphs (without sophisticated software), graphing is generally of little use in solving systems involving three variables.

We are left therefore with **substitution** or **elimination** as the primary means for solving three-equation systems. Let us work backwards and set up a three-equation system where we set $x = 2$, $y = 3$, and $z = 4$. Then we will write three equations that are valid; that is, which are consistent with the values that we have chosen for the variables:

$$x + y + z = 9$$
$$2x + y + 2z = 15$$
$$x + 2y + z = 12$$

Using the **substitution** method for solving equations, we then have

$$x + y + z = 9, \quad \textit{and so} \quad z = 9 - x - y$$

$$\textit{then } 2x + y + 2(9 - x - y) = 2x + y + 18 - 2x - 2y = 15, \textit{ and } -y = -3, \textit{ or } y = 3$$

$$\textit{and then } x + 3 + z = 9, \textit{ or } \quad x + z = 6$$

$$\textit{but } \underline{\textit{also}}\ x + 2(3) + z = 12, \textit{ or } \quad \underline{x + z = 6}$$

Therefore we can not solve these equations because we can not resolve $x + z = 6$ to say whether the answer is $1 + 5 = 6$, $2 + 4 = 6$ (the correct answer), or $3 + 3 = 6$. The problem lies in the original three equations which we chose arbitrarily. If these three equations are summed, we have

$$4x + 4y + 4z = 36, \quad \textit{or } x + y + z = 9 \ (\textit{one of the original equations})$$

The three original equations are thus not truly independent.

Let us choose three equations that are in fact independent, such as

$$x + y + z = 9$$

$$3x + y + 2z = 17$$

$$x + 3y + z = 15$$

$$\textit{which when summed give} \quad 5x + 5y + 4z = 41 \ (\textit{a new independent equation})$$

To solve these equations by the substitution method, we then have

$$x + y + z = 9, \textit{ so } z = 9 - x - y$$

$$\textit{then } 3x + y + 2(9 - x - y) = 17, \textit{ and } 3x + y + 18 - 2x - 2y = 17, \textit{ or } x - y = -1$$

$$\textit{and then } x + 3y + (9 - x - y) = 15, \textit{ and } x + 3y + 9 - x - y = 15, \textit{ or } 2y = 6, \underline{y = 3}$$

$$\textit{finally, } x - y(= 3) = -1, \textit{ so } \underline{x = 2}, \textit{ and } x(= 2) + y(= 3) + z = 9, \textit{ so } \underline{z = 4}$$

The substitution method for solving linear equations is straight forward, but it becomes laborious when the equations are more complex, or greater in number.

The **elimination** technique tends to make the calculations simpler since the mathematical operations are performed on the *total* equation, and not individual terms *within* the equation. One seeks to reduce three equations in three variables down to two equations in two unknowns, and finally to one equation in one unknown.

Let us use the three equations from above again,

$$x + y + z = 9 \qquad (1)$$

$$3x + y + 2z = 17 \qquad (2)$$

$$x + 3y + z = 15 \qquad (3)$$

The procedure is illustrated below. Say we wish to eliminate x first. We would subtract eqn (1) from eqn (3), which gives $2y = 6$ or $y = 3$. Next we would multiply eqn (1) by 3, and

$$\begin{array}{ll} (3) & x + 3y + z = 15 \\ -(1) & \underline{-(x + y + z = 9)} \\ & 2y = 6 \textit{ so } \underline{y = 3} \end{array}$$

and then

$$\begin{array}{ll} 3(1) & 3(x + y + z = 9) \\ -(2) & \underline{-(3x + y + 2x = 17)} \\ & 2y + z = 10 \textit{ so } 10 - 2(3) = \underline{z = 4} \end{array}$$

and, finally

$$x + y + z = 9, \textit{ or } 9 - 3 - 4 = \underline{x = 2}$$

subtract eqn (2) fro[illegible] ct, wh[illegible] $2y$ [illegible] en substituting $2y = 6$ into $2y + z = 10$, we fin[illegible] And [illegible] 3 an[illegible] eqn (1) shows that $x = 2$. The values for x, y and [illegible] he th[illegible] ns [illegible] can thus be found in a fairly easy and *systemati*[illegible]

Some stude[illegible] iculty [illegible] im[illegible] dure because they believe you can only subtract ([illegible] ation [illegible]) th[illegible] *just above*. But there is no significance in the [illegible] erent eq[illegible] e li[illegible] aps we should say "stacked." You can subtract o[illegible] of eq[illegible] thou[illegible] where they lay in the stack of equations.

The elimina[illegible] re ca[illegible] to a[illegible] of equations. But it becomes more tedious when [illegible] or m[illegible] equal[illegible] solved. The need to simplify the elimination pro[illegible] origi[illegible] force[illegible] velopment of **matrix algebra**.

Chapter 6.

Matrix Algebra

Some mathematicians consider matrix algebra to have "...a major role as one of the most important undergraduate mathematics topics." Matrix algebra is also the mathematical basis for the calculation of arrays of data in the "spreadsheet" computer programs widely used in business, science and engineering. But it represents a stumbling block for beginning math students. The familiar x, y, and z disappear, and are replaced by a new set of symbols that seem obscure and mysterious.

The idea behind matrix algebra has been known for more than 2000 years. But it was largely developed in Europe by **C. F. Gauss** (*1777-1855*). In his study of the orbit of a certain asteroid, Gauss used matrix algebra to solve a system consisting of *six* linear equations and six unknowns. The Gauss procedure allowed the *elimination* of equations from a system of linear equations (pg. 65), and the method is now known as **Gaussian elimination.**

Consider a general linear equation such as

$$Ax + By + Cz = D$$

Now assume that we wish go make the Ax term equal to 0. We can do this by causing *either* the coefficient A, or the variable x, to go to zero, that is,

$$Ax = 0, \quad \textit{if } \underline{\textit{either}} \ A \textit{ or } x = 0$$

This is the **basis** of **matrix algebra**.

In a set of linear equations, the values of the variables are unknown, but we *do* know the value of each coefficient. Therefore we manipulate the *individual* linear equations in such a way as to cause the coefficients of the various terms to go to zero. To illustrate, let us consider our three original equations where $x = 2$, $y = 3$ and $z = 4$.

$$x + y + z = 9 \qquad (1)$$

$$3x + y + 2z = 17 \qquad (2)$$

$$x + 3y + z = 15 \qquad (3)$$

If we multiply eqn (1) by 3, and then subtract eqn (2) from that product, we produce a new equation, $0x + 2y + z = 10$. Students often found it difficult to accept that equations can be treated in this way, but it is really only a manifestation of the *power* of the equal sign. Let

$$x + y + z = 9, \textit{ then it must be that } 3(x + y + z) = 3(9)$$

and subtracting eqn (2) from 3 × eqn (1) gives,

$$\begin{array}{rr} 3 \times \textit{eqn}\ (1) & 3x + 3y + 3z = 27 \\ \textit{minus eqn}\ (2) & \underline{-3x + -y + -2z = -17} \\ & 0x + 2y + z = 10 \end{array}$$

Note that the value of the variables *never* changes as the equations are manipulated. This can be checked at any point in the procedure by substituting in the values of the variables; i.e., $0(2) + 2(3) + 4 = 10$.

We now substitute this new equation for the original eqn (1). This is an **important point** for students to understand. To solve for three unknowns, one must have three *valid* equations. *But* these equations do not have to be the "original" equations.

We next subtract 3 × eqn (3) from eqn (2) to give

$$\begin{array}{rr} \textit{eqn}\ (2) & 3x + y + 2z = 17 \\ \textit{minus}\ 3 \times \textit{eqn}\ (3) & \underline{-3(x + 3y + z) = -3(15)} \\ & 0x - 8y - z = -28 \end{array}$$

If this last equation is substituted for the original eqn (2), our new set of three linear equations now consists of

$$\begin{aligned} 0x + 2y + z &= 10 \\ 0x - 8y - z &= -28 \\ x + 3y + z &= 15 \end{aligned}$$

To make the y coefficient in eqn (2) equal to zero, we add 4 × eqn (1) to eqn (2), which gives

$$\begin{array}{rr} 4 \times \textit{eqn}\ (1) & 4(0)x + 8y + 4z = 40 \\ \textit{plus eqn}\ (2) & \underline{0x - 8y - z = -28} \\ & 3z = 12 \\ & \textit{or}\ \underline{z = 4}\ (\textit{its correct value}) \end{array}$$

There are two "*y-z*" equations in our last set of equations, $0x+2y+z=10$ and $0x-8y-z=-28$. And substituting $z=4$ into either equation gives us our known value for y, i.e., $y = 3$. But we now have *4* valid equations, and we only need *3* to solve for x, y and z. We can therefore discard one of the "y-z" equations, which we choose to be $0x-8y-z=-28$.

The remaining three equations can then be *rearranged* (n.b., the equations can be listed in *any* order) to produce the so-called "**Gaussian elimination formation**" of three linear equations where the zero coefficients for x and y are located in the lower left portion of the formation, i.e.,

$$\begin{aligned} x + 3y + z &= 15 \\ 0x + 2y + z &= 10 \\ 0x + 0y + 3z &= 12 \end{aligned}$$

The Gaussian formation allows for convenient "back-calculation", which starting with $3z = 12$, yields the $x = 2$, $y = 3$, and $z = 4$ values in a systematic and easy way.

Later **W. Jordan** (*1842 - 1899*) recognized that the equations could be manipulated further to give a final "reduced" form of the three equations which would be

$$\begin{aligned} x + 0 + 0 &= 2 \\ 0 + y + 0 &= 3 \\ 0 + 0 + z &= 4 \end{aligned}$$

This form has the advantage that the variable values can be read directly without back-calculation. This new type of elimination is referred to as the **Gauss-Jordan elimination**.

To convert the Gauss elimination above to a Gauss-Jordan elimination, we first divide eqn (3) by *1/3* to yield a new eqn (3),

$$\frac{(0 + 0 + 3z)}{3} = \frac{12}{3}, \quad \textit{or } \underline{0 + 0 + z = 4}$$

We then subtract the new eqn (3) from eqn (2), and divide by *2* giving

$$(0 + 2y + z - 10) - (0 + 0 + z - 4) = \frac{(0 + 2y + 0 - 6)}{\textit{divide by } 2},$$

$$\textit{gives } \underline{0 + y + 0 = 3}$$

This becomes the new eqn (2). To reduce the original eqn (1), $x + 3y + z = 15$, we subtract 3 × eqn (2) from old eqn (1), giving

$$(x + 3y + z = 15) - 3(0 + y + 0 = 3) = x + 0 + z = 6$$

We then subtract eqn (3) from this equation to give the final eqn (1),

$$(x + 0 + z = 6) - (0 + 0 + z = 4) = \underline{x + 0 + 0 = 2}$$

When these "reduced"equations are collected, they yield the example of the Gauss-Jordan elimination form given above.

In these cal[illegible] have [illegible] two [illegible] we multiplied or divided an *entire* equation by [illegible] and (2) [illegible] added [illegible] ions, or subtracted one equation from another. These [illegible] arithm[illegible] ula[illegible] matrix algebra the equations would be referred to [illegible] and the [illegible] that [illegible] ned on them would be known as **row operations**.

The Matri[illegible]

Just the way [illegible] algebra [illegible] is a [illegible] students. Consider the equation: $ax + by$ [illegible] eyes [illegible] ter[illegible] n the variables *x, y,* and *z*. But it is actually the *co*[illegible] *con*[illegible] de[illegible] solution to the equation.

In matrix al[illegible] concentr[illegible] the [illegible] s and constant of the equation. This saves labor. I[illegible] exam[illegible] dow[illegible] icients and equation values of six unknowns in si[illegible] han to [illegible] ll six [illegible] in full. Our three equations can be written in m[illegible] as

[illegible: matrix with column heading *y* and rows labelled eqn 1, eqn 2, eqn 3; visible entries 1, 1, 3]

In this initial versio[illegible] included [illegible] labels, e[illegible] *eqn 1..eqn 2..eqn 3*, to help us remember what the [illegible] esents. [illegible] this pla[illegible] the mathematical calculations, however, and for m[illegible] we ca[illegible] the [illegible] mply as

[illegible: augmented matrix; visible entries 1 | 9, 2 | 17, 1 | 13]

The nine-term matr[illegible] t side [illegible] or [illegible] is known as the **coefficient matrix**. The single [illegible] repla[illegible] al [illegible] the three terms on the right side (the values of [illegible] equ[illegible] inclu[illegible] al twelve-term matrix is referred to as an **au**[illegible] **atrix**.

The matrix [illegible] **elem**[illegible] brackets. The elements are identified by subsc[illegible] whi[illegible] the [illegible] and **column** (second) in which they occur. The ma[illegible] ually [illegible] such as **A**, **B**, etc., or other capital letters which [illegible] the [illegible] example,

$$A = \begin{bmatrix} a_{11} & a_{12} & a_{13} \\ a_{21} & a_{22} & a_{23} \\ a_{31} & a_{32} & a_{33} \end{bmatrix} \qquad B = \begin{bmatrix} b_{11} & b_{12} & b_{13} & b_{14} \\ b_{21} & b_{22} & b_{23} & b_{24} \end{bmatrix} \qquad C = \begin{bmatrix} c_{11} \\ c_{12} \\ c_{13} \end{bmatrix}$$

(3 x 3) *(2 x 4)* *(1 x 3)*

Notice that the rows go *across* and the columns run *down*. Remember we set in rows, but march in columns. The matrix **A** is termed a "*3 x 3*" matrix (the number of rows is always given first) since it has three rows and three columns. The matrix **B** is a "*2 x 4*" matrix since it has two rows and four columns. Matrix **C** is a "*1 x 3*" matrix, but it is also known as a **column matrix**. If a matrix consists only of a *single* row, no matter how many columns, it is called a **row matrix**. Both column and row matrices may be referred to, in certain contexts, as **vector matrices**.

The elements of matrix **A** are generally written as *a*, those of matrix **B** as *b*, those of matrix **C** as *c*, and so on. This tells us which matrix served as the origin for the element. The position of the element in the matrix is given as a_{ij}, where the subscript *i* is the number of the row, and subscript *j* is the number of the column. For example, the element a_{21} is the element that lies in matrix **A** in row *two* and column *one*.

Beginning students often experience "eye shock" when they first see an expression such as

$$|A| = a_{11}a_{22} - a_{21}a_{12}$$

But this expression just describes the calculation of the **determinant** (see below) of a *2 x 2* matrix. One goes to the original matrix, extract the values for the different elements from that matrix, and then perform the math operations called for, e.g.,

$$A = \begin{bmatrix} 3 & -1 \\ 2 & 3 \end{bmatrix} = \begin{bmatrix} a_{11} & a_{12} \\ a_{21} & a_{22} \end{bmatrix}$$

$$\textit{and so } |A| \textit{ (the determinant)} = 3(3) - (-1)2 = 11$$

There are several specialized terms that are used in referring to matrix systems. Here are two. The formation produced by a Gaussian elimination is said to be in **row echelon form**. The form resulting from a Gauss-Jordan elimination is termed a **reduced row echelon form**.

$$\begin{bmatrix} a_{11} & a_{12} & a_{13} \\ 0 & a_{22} & a_{23} \\ 0 & 0 & a_{33} \end{bmatrix} \qquad \begin{bmatrix} a_{11} & 0 & 0 \\ 0 & a_{22} & 0 \\ 0 & 0 & a_{33} \end{bmatrix}$$

row echelon form *reduced row echelon form*

The word *echelon* means a formation where the succeeding elements fall behind, and to the left or right, of the original element. Flying geese and fighter aircraft fly in echelon form. The diagonal line from a_{11} to a_{33} is known as the **diagonal**, and the matrix that is reduced to just the diagonal, with all other elements being equal to zero, is said to be **diagonalized.**

Arithmetic operations with matrices

We will look only at the elementary operations of matrix algebra. The student should recognize, however, that these are the basic steps upon which the more sophisticated matrix calculations are built.

Addition-subtraction. Note first that *only* matrices of the **same size** can be added or subtracted. A *"2 x 3"* matrix can not be added or subtracted from a *"4 x 3"* matrix, for example. If the matrices are of the same size, addition or subtraction is just a matter of adding or subtracting the corresponding elements of the two matrices. That is,

Addition

$$\begin{bmatrix} 3 & 5 \\ 2 & 7 \end{bmatrix} + \begin{bmatrix} 2 & -4 \\ 3 & 4 \end{bmatrix} = \begin{bmatrix} (3+2) & (5+(-4)) \\ (2+3) & (4+7) \end{bmatrix} = \begin{bmatrix} 5 & 1 \\ 5 & 11 \end{bmatrix}$$

Subtraction

$$\begin{bmatrix} 4 & 2 \\ 5 & 3 \end{bmatrix} - \begin{bmatrix} 2 & 4 \\ 2 & 1 \end{bmatrix} = \begin{bmatrix} (4-2) & (2-4) \\ (5-2) & (3-1) \end{bmatrix} = \begin{bmatrix} 2 & -2 \\ 3 & 2 \end{bmatrix}$$

The adding or subtracting of matrices is therefore straightforward.

Multiplication. A matrix may be multiplied by either a **scalar** (i.e., a real number) or another matrix. Multiplication of a matrix by a scalar is simple. Each element of the matrix is just multiplied by the scalar, i.e.,

$$5 \; x \begin{bmatrix} 3 & 6 \\ 2 & 4 \end{bmatrix} = \begin{bmatrix} 5\cdot 3 & 5\cdot 6 \\ 5\cdot 2 & 5\cdot 4 \end{bmatrix} = \begin{bmatrix} 15 & 30 \\ 10 & 20 \end{bmatrix}$$

The **multiplication of two matrices** follows a procedure that at first appears strange and arbitrary. It is called **row x column multiplication**. Here we multiply the matrix **A** by matrix **B** to yield matrix **C**.

$$\boldsymbol{A} \begin{bmatrix} 2 & 4 \\ 5 & 3 \\ 4 & 5 \end{bmatrix} \times \boldsymbol{B} \begin{bmatrix} 3 & 8 & 0 & 2 \\ 3 & 4 & 3 & 4 \end{bmatrix} = \boldsymbol{C}$$

$$\boldsymbol{C} = \begin{bmatrix} 2x3+4x3=18 & 2x8+4x4=32 & 2x0+4x3=12 & 2x2+4x2=12 \\ 5x3+3x3=24 & 5x8+3x4=52 & 5x0+3x3=9 & 5x2+3x4=22 \\ 4x3+5x3=27 & 4x8+5x4=52 & 4x0+5x3=15 & 4x2+5x4=28 \end{bmatrix}$$

$$\boldsymbol{C} = \begin{bmatrix} 18 & 32 & 12 & 12 \\ 24 & 52 & 9 & 22 \\ 27 & 52 & 15 & 28 \end{bmatrix}$$

We can summarize the matrix multiplication procedures in generalized form as,

$$\boldsymbol{A} = \begin{bmatrix} a_{11} & a_{12} \\ a_{21} & a_{22} \\ a_{31} & a_{32} \end{bmatrix} \times \boldsymbol{B} = \begin{bmatrix} b_{11} & b_{12} & b_{13} & b_{14} \\ b_{21} & b_{22} & b_{23} & b_{24} \end{bmatrix} = \boldsymbol{C}$$

and Row 1 *of* ***C*** *is* $c_{11} = (a_{11}b_{11} + a_{12}b_{21})$

$c_{12} = (a_{11}b_{12} + a_{12}b_{22})$

and so on...

Notice that we multiply the *first* element in *Row* 1 of **A** × the *first* element of *Column* 1 of **B**, then the *second* element of Row 1 in **A** × the *second* element of Column 1 of **B**, and finally sum the two products to yield the *first* element of Row 1 of **C**, i.e., c_{11}.

Repeating th [illegible] with th [illegible] completes the four elements in Row 1 of **C**, that is [illegible] d c_1 [illegible] of **A** and repeating the process with the four Colu [illegible] es the [illegible] ., c_{21}, c_{22}, c_{23} and c_{24}. And finally, doing the s [illegible] with [illegible] and [illegible] lumns of **B** yields the four elements of Row 3 [illegible] c_{32}, [illegible] Th [illegible] the matrix, **C**.

Example o [illegible] ltipli [illegible] lans to build *10* houses of Type A, *20* houses of Ty [illegible] ouses [illegible] pes of houses will be offered (in different numbers) i [illegible] -- m [illegible] adi [illegible] builder needs to know how much material he w [illegible]

This problem [illegible] ted to [illegible] ebra [illegible] stablish the matrix, **TH**, which lists the total numb [illegible] nd i [illegible] style, viz.,

[illegible]

Type [illegible]

Type [illegible] *Tot* [illegible] *of Houses* (***TH***)

Type [illegible]

Then we est [illegible] nd m [illegible] , wh [illegible] he amounts of building materials needed (in [illegible] its) for [illegible] diff [illegible] of houses,

	[illegible]	*Lum* [illegible]	[illegible]	
Modern	[illegible]	10	[illegible]	*Materials/House* (***M/H***)
Traditional	[illegible]	10	[illegible]	

If we then d [illegible] ultipl [illegible] he " [illegible] " matrix, **TH**, by the "materials per house" matrix, [illegible] tain th [illegible] terial [illegible] **TM**, as shown here

$$\boldsymbol{TH}\begin{bmatrix} 3 & 7 \\ 5 & 15 \\ 10 & 5 \end{bmatrix} \times \boldsymbol{M/H}\begin{bmatrix} 5 & 10 & 0 & 5 \\ 6 & 10 & 8 & 2 \end{bmatrix} = \boldsymbol{TM}$$

$$\boldsymbol{TM} = \begin{bmatrix} 3x5+7x6=57 & 3x10+7x10=100 & 3x0+7x8=56 & 3x5+7x2=29 \\ 5x5+15x6=115 & 5x10+15x10=200 & 5x0+15x8=120 & 5x5+15x2=55 \\ 10x5+5x6=80 & 10x10+10x5=150 & 10x0+5x8=40 & 10x5+5x2=60 \end{bmatrix}$$

$$\boldsymbol{TM} = \begin{bmatrix} 57 & 100 & 56 & 29 \\ 115 & 200 & 120 & 55 \\ 80 & 150 & 40 & 60 \end{bmatrix}$$

The matrix, **TM**, is "bare bones" and not easy to visualize. However, if we add the title and labels to the matrix, then it becomes understandable, i.e.,

Total Materials by Matrix Algebra

	Concrete	*Lumber*	*Brick*	*Roofing*
Type A	57	100	56	29
Type B	115	200	120	55
Type C	80	150	40	60

This matrix gives the total material per type of house. To find the materials for the *total* building project, the columns must be summed, e.g., the concrete needs are *57 + 115 + 80* or *252 units* in total, and so on.

We can check this result by "regular" math. The Type A houses consist of *3* modern and *7* traditional style houses. And the material requirements for each style of house are known. Multiplying the number of each style of house by the specific material requirements for that house, and summing the two products, yields the total amount of material required, i.e.,

Table of Materials for Type A Houses

		Concrete	*Lumber*	*Brick*	*Roofing*
3 mod. *houses*	= 3(5 10 0 5) =	15	30	0	15
7 *trad. houses*	= 7(6 10 8 2) =	42	70	56	14
Total material		57	100	56	29

This is equivalent to Row 1 for Type A houses in the matrix table of materials. If we repeat the same calculations with the Type B and Type C houses, we reproduce the values of the Matrix Table of Materials, which verifies that matrix algebra gives the correct answer:

Total Materials by Regular Math

	Concrete	*Lumber*	*Brick*	*Roofing*
Type A	57	100	56	29
Type B	115	200	120	55
Type C	80	150	40	60

Notice that matrix algebra and regular math differ only in the "direction" of the math operations. In matrix algebra, we multiply the number of A-mod houses (3) by A-mod concrete (5) (equals *15*), and number of A-trad houses (7) by A-trad concrete (6) (equals *42*) and add the two products *horizontally* to give element c_{11} (equals *57*) of matrix **C** directly. In regular math, we multiply A-mod houses by A-mod concrete (equals *15*) and A-trad houses by A-trad concrete (equals *42*) and then sum the two products (equals *57*) *vertically* to give the total concrete need for Type A houses.

The advantage to matrix algebra is that it is more direct, and when used in spreadsheet software, presents detailed answers in a matrix form that can be displayed in an easy-to-read style on the computer screen or printout.

Two points are important in matrix **multiplication**. First, the number of *columns* in the first matrix must be equal to the number of *rows* in the second matrix. Look at the matrix multiplication of Total Houses (**TH**) times Materials/House (**T/M**) to determine Total Material (**TM**).

$$\boldsymbol{TH_{3 \times 2} \times M/H_{2 \times 4} = TM_{3 \times 4}}$$

We have *two columns* in the matrix **TH** indicating the number of modern and traditional style of houses. We have *two rows* in the matrix **M/H**, which give the materials required per house of the modern or traditional style. We can multiply these. But, if there was an *extra* style of house (say Cape Cod) *without* an extra list of materials per house of that style, the matrices could *not* be multiplied. That is, we could not determine the *total amount* of material needed for the Cape Cod style of houses.

Note also that Total Materials (**TM**) is a *3 x 4* matrix, and that the number of *rows* in **TM** is equal to the number of *rows* in **TH**, while the number of *columns* in **TM** is equal to the number of *columns* in **M/H**. This follows from the nature of matrix multiplication as we see evidenced here,

$$A \begin{bmatrix} a_{11} & a_{12} \\ a_{21} & a_{22} \\ a_{31} & a_{32} \end{bmatrix} \times B \begin{bmatrix} b_{11} & b_{12} & b_{13} & b_{14} \\ b_{21} & b_{22} & b_{23} & b_{24} \end{bmatrix} = C \begin{bmatrix} c_{11} & c_{12} & c_{13} & c_{14} \\ c_{21} & c_{22} & c_{23} & c_{24} \\ c_{31} & c_{32} & c_{33} & c_{34} \end{bmatrix}$$

The second point is that matrix multiplication is **not communicative**; that is, the multiplication of matrix **A** by matrix **B** does *not* give the same answer as the multiplication of matrix **B** by matrix **A**. Let us look at two square matrices, **A** and **B**, which we have constructed from the first four elements of the matrices, **TH** and **M/H**.

$$A = \begin{bmatrix} 3 & 7 \\ 5 & 15 \end{bmatrix} \quad B = \begin{bmatrix} 5 & 10 \\ 6 & 10 \end{bmatrix} \quad and\ AB = \begin{bmatrix} 57 & 100 \\ 115 & 200 \end{bmatrix}$$

$$but\ BA = B \begin{bmatrix} 5 & 10 \\ 6 & 10 \end{bmatrix} \times A \begin{bmatrix} 3 & 7 \\ 5 & 15 \end{bmatrix} = \begin{bmatrix} 65 & 185 \\ 88 & 192 \end{bmatrix} \quad and\ so\ AB \neq BA$$

To be equal, matrices must be of the same size, and each element must be identical which is *not* the case here. So **AB** is clearly not identical to **BA**.

This is sometimes confusing to students. But it is actually a matter of semantics. We say that we "multiply matrices," but in reality the elements are found by a combination of multiplication and addition.

$$c_{11} = a_{11}b_{11} + a_{12}b_{21}$$

This gives that, for **AB**, c_{11} is equal to $3x5 + 7x6 = 57$, but for **BA**, c_{11} is equal to $5x3 + 10x5 = 65$. Therefore **AB** and **BA** can not be the same.

We got the right answer in our "house building" exercise because the first column of matrix **TH** (the numbers of modern houses) was multiplied by the first row in matrix **M/H** (materials needed for a modern house). If we had made the first row of matrix **M/H** the materials required for a *traditional* house, our answer would have been wrong.

Division. It is not possible to "reverse" multiplication to give division, so there is *no division* of a matrix *by a matrix* in matrix algebra. But a matrix can be divided by a **scalar** (i.e., a real number). If the builder chose to build only ½ the number of houses, for example, he would simply divide the matrix **TM** by 2 to find the new amounts of materials needed.

$$\frac{1}{2} \times TM \begin{bmatrix} 57 & 1[illegible] & [illegible] & [illegible] \\ 115 & 2[illegible] & [illegible] & [illegible] \\ 80 & 1[illegible] & [illegible] & [illegible] \end{bmatrix} = \begin{bmatrix} [illegible] & [illegible] & \frac{55}{2} \\ [illegible] & [illegible] & \frac{120}{2} \\ [illegible] & [illegible] & \frac{40}{2} \end{bmatrix}$$

$$\text{[illegible]w } TM \begin{bmatrix} 23.5 & 50 & 28 & 14.5 \\ 57.5 & 100 & 60 & 27.5 \\ 40 & 75 & 20 & 30 \end{bmatrix}$$

Other op[illegible] ith m[illegible]

While we ca[illegible] trix-b[illegible] visi[illegible] o a version of division by the use of the **inverse** [illegible] com[illegible] ith [illegible] **matrix**. Examples of identity matrices are

$$[illegible] \quad \begin{bmatrix} [illegible] & [illegible] & 0 \\ [illegible] & [illegible] & 0 \\ [illegible] & [illegible] & 1 \end{bmatrix} \quad \begin{bmatrix} 1 & 0 & 0 & 0 \\ 0 & 1 & 0 & 0 \\ 0 & 0 & 1 & 0 \\ 0 & 0 & 0 & 1 \end{bmatrix}$$

Identity matrices ar[illegible] by **I**, [illegible] th[illegible] ndicates the order of the matrix. The elements of th[illegible] low[illegible] iagonal [illegible] of *1's* while all of the rest of the elements are *ze*[illegible]

Identity ma[illegible] ways [illegible] is, [illegible] ays *2 x 2*, *3 x 3*, *4 x 4*, etc. in size. Their importa[illegible] in the [illegible] iscussi[illegible] ny matrix, **A**, multiplied by its *inverse*, which we [illegible] $^{-1}$, ret[illegible] ti[illegible]

$$[illegible](A^{-1}) = [illegible] = I$$

Note that A^{-1} does [illegible] eciprc[illegible] as i[illegible] real numbers. And also that A^{-1} is not found by [illegible] ipro[illegible] elem[illegible] natrix **A**.

To find the i[illegible] matri[illegible]

$$\begin{bmatrix} 4 & [illegible] \\ -3 & [illegible] \end{bmatrix}$$

we write out the re[illegible]

$$A(A^{-1}) = I \quad or \quad A\begin{bmatrix} 4 & 2 \\ -3 & 6 \end{bmatrix} \times A^{-1}\begin{bmatrix} a_{11} & a_{12} \\ a_{21} & a_{22} \end{bmatrix} = I_2\begin{bmatrix} 1 & 0 \\ 0 & 1 \end{bmatrix}$$

And we see that we need to determine the elements in the *inverse matrix* $\mathbf{A}^{-1}$ which would make $\mathbf{A} \, x \, \mathbf{A}^{-1} = \mathbf{I}_2$.

There is a shortcut for determining the inverse of a *2 x 2* matrix. But be aware that this shortcut does *not* work for matrices larger than *2 x 2*. To use this shortcut, one switches the elements of the main diagonal

$$\begin{bmatrix} 4 & 2 \\ -3 & 6 \end{bmatrix} \rightarrow \begin{bmatrix} 6 & 2 \\ -3 & 4 \end{bmatrix}$$

and then takes the opposite of the other two elements (i.e., change their signs)

$$\begin{bmatrix} 6 & 2 \\ -3 & 4 \end{bmatrix} \rightarrow \begin{bmatrix} 6 & -2 \\ 3 & 4 \end{bmatrix}$$

The **determinant** (see below) of this array is given by $6 \, x \, 4 - (-2) \, x \, 3 = 30$. The determinant is then used as the denominator of the elements in the matrix $\mathbf{A}^{-1}$. This yields the matrix

$$A^{-1} = \begin{bmatrix} \frac{6}{30} & \frac{-2}{30} \\ \frac{3}{30} & \frac{4}{30} \end{bmatrix}$$

If it happened that element a_{12} was *8* and not *-2*, the determinant would be $6 \, x \, 4 - (8 \, x \, 3) = 0$. And this particular matrix would then have *no* inverse, since we can not divide by *zero*.

We can verify that the inverse $\mathbf{A}^{-1}$ is correct by multiplying the two matrices

$$A \begin{bmatrix} 4 & 2 \\ -3 & 6 \end{bmatrix} \times A^{-1} \begin{bmatrix} \frac{6}{30} & \frac{-2}{30} \\ \frac{3}{30} & \frac{4}{30} \end{bmatrix} = (?)\ I_2 \begin{bmatrix} 1 & 0 \\ 0 & 1 \end{bmatrix}$$ *yes, see below*

$$i_{11} = \frac{4\times 6}{30} + \frac{2\times 3}{30} = \frac{30}{30} = 1$$
$$i_{12} = \frac{4\times(-2)}{30} + \frac{2\times 4}{30} = 0$$
$$i_{21} = \frac{(-3)\times 6}{30} + \frac{6\times 3}{30} = 0$$
$$i_{22} = \frac{(-3)(-2)}{30} + \frac{6\times 4}{30} = 1$$

Square matrices larger than 2 x 2. One can find the inverse matrix of square matrices larger than *2 x 2* by the **augmented matrix** method. This works for all *n x n* matrices. We will demonstrate with the method using our prior three linear equations, i.e.,

$$\begin{aligned} x + y + z &= 9 \\ 3x + y + 2z &= 17 \\ x + 3y + z &= 15 \end{aligned}$$

We **adjoin** the coefficient matrix **A** of the equations with the identity matrix **I** to give the augmented matrix

$$\left[\begin{array}{ccc|ccc} 1 & 1 & 1 & 1 & 0 & 0 \\ 3 & 1 & 2 & 0 & 1 & 0 \\ 1 & 3 & 1 & 0 & 0 & 1 \end{array}\right]$$

Note that we have *not* added matrix **A** to matrix **I**. That would give

$$A \begin{bmatrix} 1 & 1 & 1 \\ 3 & 1 & 2 \\ 1 & 3 & 1 \end{bmatrix} + I \begin{bmatrix} 1 & 0 & 0 \\ 0 & 1 & 0 \\ 0 & 0 & 1 \end{bmatrix} = \begin{bmatrix} 2 & 1 & 1 \\ 3 & 2 & 2 \\ 1 & 3 & 2 \end{bmatrix}$$

where the elements of the two matrices are simply added together.

Rather we have “stuck ” matrix **I** onto matrix **A** to produce a new matrix that is *3 x 6* in size. It is a **property of matrices** that they can broken down into smaller matrices, or added together to produce larger matrices, for mathematical purposes.

We will use **row operations** to find the inverse matrix $\mathbf{A}^{-1}$ of the matrix **A**, i.e.,

$$\begin{array}{cccccc} 1 & 1 & 1 & 1 & 0 & 0 \\ 3 & 1 & 2 & 0 & 1 & 0 \\ 1 & 3 & 1 & 0 & 0 & 1 \end{array} \; row2 - 3\times row1 \rightarrow \; \begin{array}{cccccc} 1 & 1 & 1 & 1 & 0 & 0 \\ 0 & -2 & -1 & -3 & 1 & 0 \\ 1 & 3 & 1 & 0 & 0 & 1 \end{array} \; row1 - row3 \rightarrow \; \begin{array}{cccccc} 1 & 1 & 1 & 1 & 0 & 0 \\ 0 & -2 & -1 & -3 & 1 & 0 \\ 0 & -2 & 0 & 1 & 0 & -1 \end{array}$$

$$row2 - row3 \rightarrow \; \begin{array}{cccccc} 1 & 1 & 1 & 1 & 0 & 0 \\ 0 & 2 & -1 & -3 & 1 & 0 \\ 0 & 0 & -1 & -4 & 1 & 1 \end{array} \; row2 - row3 \rightarrow \; \begin{array}{cccccc} 1 & 1 & 1 & 1 & 0 & 0 \\ 0 & -2 & 0 & 1 & 0 & -1 \\ 0 & 0 & -1 & -4 & 1 & 1 \end{array} \; -1/2 row2 \rightarrow$$

$$\begin{array}{cccccc} 1 & 1 & 1 & 1 & 0 & 0 \\ 0 & 1 & 0 & -1/2 & 0 & 1/2 \\ 0 & 0 & 1 & 4 & -1 & -1 \end{array} \; row1 - row3 \rightarrow \; \begin{array}{cccccc} 1 & 1 & 0 & -3 & 1 & 1 \\ 0 & 1 & 0 & \frac{-1}{2} & 0 & \frac{1}{2} \\ 0 & 0 & 1 & 4 & 0 & -1 \end{array} \; row1 - row2 \rightarrow \; \begin{array}{cccccc} 1 & 0 & 0 & \frac{-5}{2} & 1 & \frac{1}{2} \\ 0 & 1 & 0 & \frac{-1}{2} & 0 & \frac{1}{2} \\ 0 & 0 & 1 & 4 & -1 & -1 \end{array}$$

At the end of *7* row operations, we have converted the original adjoined matrix to,

$$\left[\begin{array}{ccc|ccc} 1 & 0 & 0 & \frac{-5}{2} & 1 & \frac{1}{2} \\ 0 & 1 & 0 & \frac{-1}{2} & 0 & \frac{1}{2} \\ 0 & 0 & 1 & 4 & -1 & -1 \end{array}\right] \quad or \;\; IA^{-1}, \;\; and\; so \;\; A^{-1} = \begin{bmatrix} \frac{-5}{2} & 1 & \frac{1}{2} \\ \frac{-1}{2} & 0 & \frac{1}{2} \\ 4 & -1 & -1 \end{bmatrix}$$

and we can “unadjoin” the identity matrix, **I**, from $\mathbf{IA}^{-1}$ to yield the inverse matrix, $\mathbf{A}^{-1}$.

One of the main uses of the inverse matrix $\mathbf{A}^{-1}$ is to solve a system of linear equations. We can write a system of linear equations in matrix terms as **A** (the coefficients matrix) × **X** (the variables matrix) = **B** (the equation values); that is, **AX** = **B**.

If we multiply the matrix **A** (coefficient matrix) by **X** (variables matrix), we regenerate the system of linear equations. We can verify this with our three known linear equations, where if we do a row × column multiplication of matrix **A** × matrix **X**, we return the original equations,

$$\begin{bmatrix} [illegible] & 1 \\ [illegible] & 2 \\ [illegible] & 1 \end{bmatrix} = \begin{bmatrix} 1x \; [illegible] \; 1z \\ 3x \; [illegible] \; 2z \\ 1x \; [illegible] \; 1z \end{bmatrix}$$

Furthermor [illegible] (whic [illegible] he values of the individual equations), then **X** [illegible] vever, [illegible] di [illegible] natrix **A**. But we can do the equivalent of divisi [illegible] ying **B** [illegible] whic [illegible] rse matrix of matrix **A**. A row × column multiplic [illegible] × **B** g [illegible] umn [illegible] hich lists the values of the variables, *x*, *y* and [illegible]

$$\begin{bmatrix} \frac{-5}{2} & [illegible] \\ \frac{-1}{2} & [illegible] \\ 4 & [illegible] \end{bmatrix} [illegible] B \begin{bmatrix} [illegible] \\ [illegible] \\ [illegible] \end{bmatrix} = \begin{matrix} x \\ y \\ z \end{matrix}$$

The details of the r [illegible] multi [illegible] are [illegible] + (1)17 +(1/2)15 = 2, *y* = (-1/2)9 − (0)17 [illegible] 3, ar [illegible] 5 = 4.

The Dete [illegible]

It is difficul [illegible] tuiti [illegible] n for [illegible] **minant**. Usually the determinant is defi [illegible] g like [illegible] um [illegible] be derived from the matrix, and which reveals certa [illegible] rix's pr [illegible] Th [illegible] nt can be positive, negative, or zero. It is normally [illegible]

$$|A| \; [illegible] = [illegible]$$

In matrix algebra, [illegible] matri [illegible] r it [illegible] ertical straight lines means that we are talking abo [illegible] *nt*, a [illegible] bs [illegible] the term.

A primary [illegible] nan [illegible] ng [illegible] near equations. And we can probably obtain ou [illegible] andi [illegible] term [illegible] dying its use in this role. In the 1700's, **G. Crame** [illegible] deve [illegible] mer [illegible] olving two linear equations. Consider the linear [illegible]

$$ax + by = e$$
$$cx + dy = f$$

$$\begin{array}{lll} \textit{then } d(ax + by) = de, \textit{ or} & dax + dby = de \\ \textit{and } -b(cx + dy) = -bf, \textit{ or} & \underline{-bcx - dby = -bf} \\ \qquad \textit{giving} & (ad - bc)x = de - bf \end{array}$$

$$\textit{therefore, } x = \frac{de - bf}{ad - bc} \quad \textit{and similarly, } y = \frac{af - ce}{ad - bc}$$

Thinking in terms of matrix algebra, we see that the expression, *ad - bc*, can be obtained from the *coefficient* matrix, if we multiply the elements on the **main** diagonal and subtract from this, the product of the elements off the main diagonal. In equation form, this would be

$$\begin{bmatrix} a & b \\ c & d \end{bmatrix} \quad \textit{giving } D = \begin{vmatrix} a & b \\ c & d \end{vmatrix} = ad - bc$$

We define the result of this operation as the **determinant** of the *2x2* coefficient matrix which we call *D*.

From above, we see that, now knowing *ad - bc*, we could solve for *x* if we knew *de - bf*. And we can find *de - bf* by replacing the *x* coefficients (*a* and *b*) in the determinant matrix by the equation values (*e* and *f*), and doing the determinant calculation. This yields *de - df,* which we label D_x. Then replacing the *y* coefficients (*b* and *d*) by the equation values (*e* and *f*) in the determinant matrix, and doing the determinant calculation yields *af - ce*, or D_y, i.e.,

$$D_x = \begin{vmatrix} e & b \\ f & d \end{vmatrix} = de - bf, \textit{ and also } D_y = \begin{vmatrix} a & e \\ c & f \end{vmatrix} = af - ce$$

Comparison with the original linear equation system shows that

$$x = \frac{D_x}{D} \quad \textit{and } y = \frac{D_y}{D}$$

A "two unknowns, two (linear) equations" problem can thus be solved in an easy and systematic manner using determinants and Cramer's rule.

A *3 x 3* matrix representing three linear equations is more difficult, but it can also be solved using determinants and matrix algebra. Let us use our three known linear equations,

$$\begin{aligned} x + y + z &= 9 \\ 3x + y + 2z &= 17 \\ x + 3y + z &= 15 \end{aligned}$$

These equations can be represented by as the augmented matrix

$$\left[\begin{array}{ccc|c} 1 & 1 & 1 & 9 \\ 3 & 1 & 2 & 17 \\ 1 & 3 & 1 & 15 \end{array}\right]$$

We will use a *shortcut* that mathematicians have developed for calculating the determinants from a *3 x 3* matrix. One constructs a special matrix (for calculation purposes) by attaching the first two columns of the coefficient matrix to the right side of the coefficient matrix itself, i.e.,

$$\begin{vmatrix} 1 & 1 & 1 & 1 & 1 \\ 3 & 1 & 2 & 3 & 1 \\ 1 & 3 & 1 & 1 & 3 \end{vmatrix}$$

Then to find D, the determinant of the coefficient matrix, we take the products of the three elements in each of the *three* diagonals starting from the *upper left corner*, and sum them. This gives *(1x1x1) + (1x2x1) + (1x3x3) = 12.* We then sum the products of the three diagonals starting from the *upper right corner*. This is *(1x3x1) + (1x2x3) + (1x1x1) = 10.* This sum is *subtracted* from our first sum, or *12 - 10 = 2.* This gives *2* as the value of D, the determinant of the coefficient matrix.

To find D_x, the determinant of the *"x"* matrix, we replace the columns which represent the *x* coefficients by the equivalent values of the different equations,

$$\begin{vmatrix} 9 & 1 & 1 & 9 & 1 \\ 17 & 1 & 2 & 17 & 1 \\ 15 & 3 & 1 & 15 & 3 \end{vmatrix}$$

The sum of products of the three diagonals from the left is *(9x1x1) + (1x2x15) + (1x17x3) = 90*, and the sum of products of the diagonals from the right is *(1x17x1) + (9x2x3) + (1x1x15) = 86.* The determinant, D_x, is *90 - 86 = 4*, and *x* itself is equal to D_x / D, or *4/2 = 2.*

To find the D_y and D_z determinants, we set up the following matrices and apply the same procedures as for D_x. Notice in the D_z matrix that the equation values are substituted for only *one* column, that being the *z* coefficient column.

$$\begin{vmatrix} 1 & 9 & 1 & 1 & 9 \\ 3 & 17 & 2 & 3 & 17 \\ 1 & 15 & 1 & 1 & 15 \end{vmatrix} \qquad \begin{vmatrix} 1 & 1 & 9 & 1 & 1 \\ 3 & 1 & 17 & 3 & 1 \\ 1 & 3 & 15 & 1 & 3 \end{vmatrix}$$

$$\textit{to calculate } D_y \qquad \textit{to calculate } D_z$$

By these procedures, we find D_y to be *6*, and D_z to be *8*. Since $D = 2$, this gives $y = D_y / D = 3$, and $z = D_z / D = 4$. These are our known values for *x, y,* and *z*, respectively.

Unfortunately, these simple methods *will not work* for matrices that are larger than *3 x 3*. The student must keep this in mind.

There are however two methods that will work for *square* matrices of any size. One method is by **row operations**, and the other is **expansion using minors and cofactors**.

Row operations. In this method, the goal is to reduce the matrix into a triangular form. Once the matrix is in triangular form (i.e., with all elements to the left of the main diagonal equal to zero), the determinant is found by *simply* multiplying the elements on the main diagonal. The elements to the right of the main diagonal are not involved. Using our previous set of three linear equations to demonstrate, we perform the necessary row operations on the "coefficient" matrix array, viz.,

$$\begin{matrix} 1 & 1 & 1 \\ 3 & 1 & 2 \\ 1 & 3 & 1 \end{matrix} \; -3\times row1 + row2 \rightarrow \begin{matrix} 1 & 1 & 1 \\ 0 & -2 & -1 \\ 1 & 3 & 1 \end{matrix} \; -1\times row1 + row3 \rightarrow \begin{matrix} 1 & 1 & 1 \\ 0 & -2 & -1 \\ 0 & 2 & 0 \end{matrix} \; row2 + row3 \rightarrow \begin{matrix} 1 & 1 & 1 \\ 0 & -2 & -1 \\ 0 & 0 & -1 \end{matrix}$$

$$\textit{then, the determinant, D,} = 1 \times (-2) \times (-1) = 2$$

To find the determinant of the "z" matrix array, we substitute the equation values in the z column, and work with the resulting matrix array,

$$\begin{matrix} 1 & 1 & 9 \\ 3 & 1 & 17 \\ 1 & 3 & 15 \end{matrix} \; -3\times row1 + row2 \rightarrow \begin{matrix} 1 & 1 & 9 \\ 0 & -2 & -10 \\ 1 & 3 & 15 \end{matrix} \; row3 - row1 \rightarrow \begin{matrix} 1 & 1 & 9 \\ 0 & -2 & -10 \\ 0 & 2 & 6 \end{matrix} \; row2 + row3 \rightarrow \begin{matrix} 1 & 1 & 9 \\ 0 & -2 & -10 \\ 0 & 0 & -4 \end{matrix}$$

$$\textit{Determinant } D_z = 1 \times (-2) \times (-4) = 8, \quad \textit{and } z = \frac{D_z}{D} = \frac{8}{2} = 4$$

Therefore we [illegible] the same [illegible] for D, [illegible] variable z as in the earlier shortcut method for [illegible] es. All [illegible] sh [illegible] x, and D_y and y, are determined in exactly the same [illegible]

The "row op [illegible] method [illegible] for [illegible] ices, but it becomes tedious with larger matrices.

Expansion [illegible] **and** [illegible] may be better for the reduction of large matrices, [illegible] comp [illegible] method has the advantage that it follows a *standard* [illegible] *on* [illegible] ds only upon the position of the element in the matri [illegible] value [illegible]

Let us use [illegible] standa [illegible] three [illegible] ons, where we have enclosed the coefficient mat [illegible] lines [illegible] the [illegible] erested in its determinant,

[illegible] 1

[illegible] 2

[illegible] 1

Let us expand arou [illegible] *w*, w [illegible] this [illegible] the determinant, D,

[illegible] = + [illegible] 3 [illegible] 1 [illegible]

= + [illegible]

This is the same de [illegible] for the [illegible] matri [illegible] have found by other methods.

This is the [illegible] *expa* [illegible] beg [illegible] first element of the Row 1 which becomes the [illegible], an [illegible] is [illegible] *cluding* the row and column in which the cofactor [illegible] de [illegible] par [illegible] The minor, which is a matrix itself, is often writt [illegible] tatio [illegible] he [illegible] ipt indicates that it is the minor of the cofactor at *e* [illegible] col [illegible] sec [illegible] in the Row 1, and its minor is the matrix excludi [illegible] and [illegible] column [illegible] this minor is *negative*. The third cofactor is the [illegible] rt of [illegible] its [illegible] matrix excluding the first row and *third* column. [illegible] facto [illegible] ltiplie [illegible] spective minors, and the products summed, [illegible] ant is [illegible]

The second [illegible] gative [illegible] sign [illegible] nor alternates from one minor to the next. If we had [illegible] s, th [illegible] uld al [illegible] -, -, +, -, +,- ... The reason for the alternation of th [illegible] with [illegible] of [illegible] at occur in the possible permutations in the [illegible] rs in [illegible] the minor (see, e.g, Linear Algebra, Determin [illegible] *Online* [illegible] *s)* [illegible] how the number of possible ways of choosing a [illegible] rs de [illegible] the binomial equation (pg. 58).

One way to [illegible] whe [illegible] term is positive or negative is to include the factor *(-* [illegible] term, [illegible] and j the column number, i.e.,

$$M_{12} = 1 \times (-1)^{1+2} \times \begin{vmatrix} 3 & 2 \\ 1 & 1 \end{vmatrix} = -1 \begin{vmatrix} 3 & 2 \\ 1 & 1 \end{vmatrix}$$

Keep in mind also that it is the sign of the *minor*, and not the sign of the cofactor, that alternates from one cofactor × minor term to the next. We write the second term of the coefficient matrix expansion as

$$-1 \begin{vmatrix} 3 & 2 \\ 1 & 1 \end{vmatrix} \textit{ but actually it is } 1 \times -\begin{vmatrix} 3 & 2 \\ 1 & 1 \end{vmatrix}$$

We have determined the determinant, *D (=2)*, of the coefficient matrix by the cofactor-minor expansion process. Now to find D_x, we replace the "x" coefficient column by the equivalent equation values, and expand around the *second* row.

$$\begin{vmatrix} 9 & 1 & 1 \\ 17 & 1 & 2 \\ 15 & 3 & 1 \end{vmatrix}$$

This expansion gives

$$D_x = 9 \begin{vmatrix} 1 & 2 \\ 3 & 1 \end{vmatrix} - 17 \begin{vmatrix} 1 & 1 \\ 3 & 1 \end{vmatrix} + 15 \begin{vmatrix} 1 & 1 \\ 1 & 2 \end{vmatrix}$$

$$= 9(-5) - (17)(-2) + 15(1) = -45 + 34 + 15 = 4$$

$$x = \frac{D_x}{D} = \frac{4}{2} = 2$$

And we see that the value for *x*, as determined by the cofactor-minor expansion, is correct.

Notice that we expanded around the *first* row in the determination of D, but around the *second* row of the matrix in our determination of D_x. This was to illustrate a property of the expansion technique. You can expand around *any* row or column of the matrix (but not a diagonal), and still obtain the correct determinant. Normally one would expand around the row or column that contains the most *zero* elements, since this reduces the calculations.

While expansion using minors and cofactors is straightforward, it can be labor intensive. For example, in the solution of a *5x5* matrix, we have

one 5*x*5 *matrix* → 5 4*x*4 *matrices*
1 4*x*4 *matrix* → 4 3*x*3 *matrices*
1 3*x*3 *matrix* → 3 2*x*2 *matrices*

so we must solve 60 2*x*2 *matrices to expand a single* 5*x*5 *matrix*

For a *6x6* matrix, one would have to solve *6 x 60 or 360* each of *2x2* matrices, and for a *7x7* matrix, *7 x 360 or 2520* each of *2x2* matrices. It is easy to see why this method is best suited for computers when *n x n* becomes large.

Matrices and determinants have been studied for hundreds of years, and they have properties well beyond the few mentioned here. Matrix mathematics all began, however, with the basic ideas introduced in this chapter.

Chapter 7.

Trigonometry

Trigonometry is the study of the properties of the **angle** that is produced by the intersection of two **straight lines**. For such a small topic, it has had an enormous impact in the development of civilization. The practices of building and architecture, the surveying and mapping of continents, navigation around the world, the construction of machines and engines, modern services such as global positioning, and many others would all be impossible except for trigonometry.

The ancient people had two ways of producing a straight line. One was the tightly stretched string which, in accord with the classical definition of a straight line, "..gives the shortest distance between two points." Even today, brick layers and carpenters use a stretched line to insure that the walls and structures they are building are straight. Vertical straightness is determined by comparison with a "plumb bob," which is a weight (pointed at the lower end) that stretches a string straight down.

The other way of producing a straight line was by "line of sight." Sighting along an arrow's shaft, or through the sights of a rifle, or along the marker stakes of a highway under construction are all ways of projecting a *straight* line forward from the observer's site. Surveying in particular is based primarily upon line of sight measurements.

A major feat of surveying was achieved in 1856 when a survey of India showed the height of Mt. Everest (then unnamed) to be *8,840* meters, or *29002* feet. This figure was obtained by summing the incremental elevation changes for hundreds of surveying sites starting at sea level. Modern studies, using GPS and other sophisticated techniques, place Mt. Everest's height at *8,848-8,850* meters. The total correction is about *10* meters over a period of 150 years, which is truly remarkable.

It is possible also to generate a **circle** using just a stretched string. One end of the string is held at a rotating center point, and the other end of the stretched string (which is fitted with a stick or stylus to mark the circle) is rotated completely around this point. A modern-day student's compass, which consists of two pointed branches joined at the top by a pivot, works in essentially the same manner.

In mathematics, the string is normally rotated in the counter-clockwise direction starting at *zero* degrees. The length of the string determines the **radius** of the circle, see Fig. 7-1.

The **diameter** of the circle can be found determined by stretching a string across the circle, with the line being made to fall upon the center of the circle, as in Fig. 7-2. Notice that the length of the diameter is *twice* that of the radius.

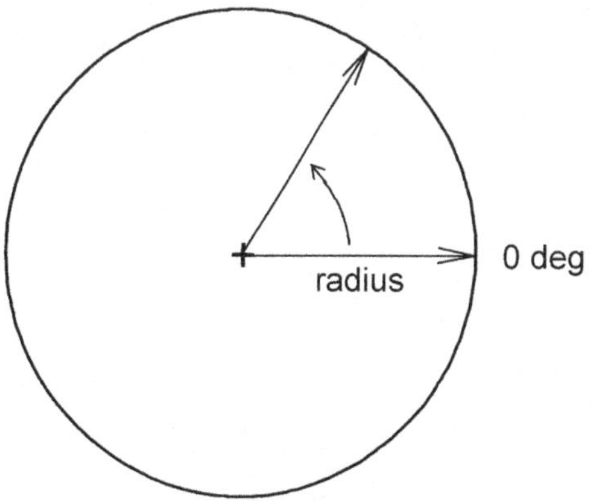

Fig. 7-1 Circle drawn by rotated string

A line **perp** ... the di... be drawn by using a s... to stri... es o... *equal* radius from t... the dia... Fig 7-2). A string is th... from th... the circle (i.e., its **peri**... gh th... on point of the two arc... center... circle, and to the opposite ... This g... line that is perpendicul... al d... which, together wi... diam... ides th... circle into four equ... s.

Our circle i... 360 d... legac... from the Babyloni... gles... the ea... quadrant are **90°** an... 0° ang... called a **right angle**. Thi... he pr... es from building prac... s wa... ls to ...licular, and their corners to be square. Using a 90... the d... lt, so ... is the "right angle."

The right an... mporta... ... of the angles of a **triangle** is a right angle, that tri... ht tri... e rig... has the property that the square of the length of the l... (the sid... from ...gle), $\mathbf{c^2}$, is equal to the sum of the square of the le... adjace... and ..., $\mathbf{c^2 = a^2 + b^2}$.

The fact tha... ^{2} for a ri... angle is ... the **Pythagorean theorem** whose proof is sho... -3. Th... s equal to $(a + b)^2$, but it is *also* equal to $c^2 + 4(1/2$ ...

$$(a + b)^2 = c^2 \; ...$$

$$a^2 + 2\,ab + \; ... \; + 2\,ab$$

$$a^2 + \; ...$$

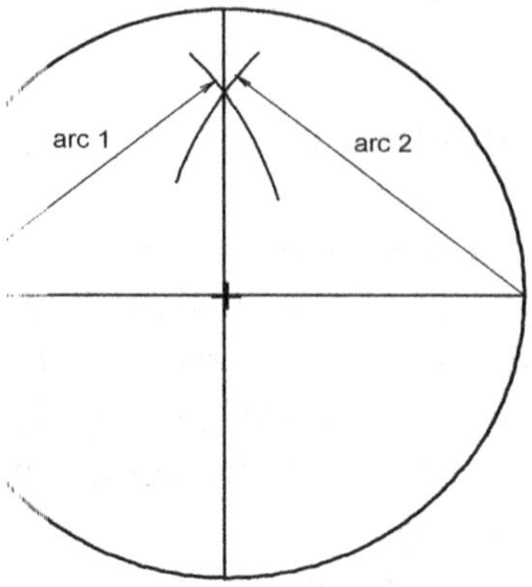

...2 Circle divided into quadrants

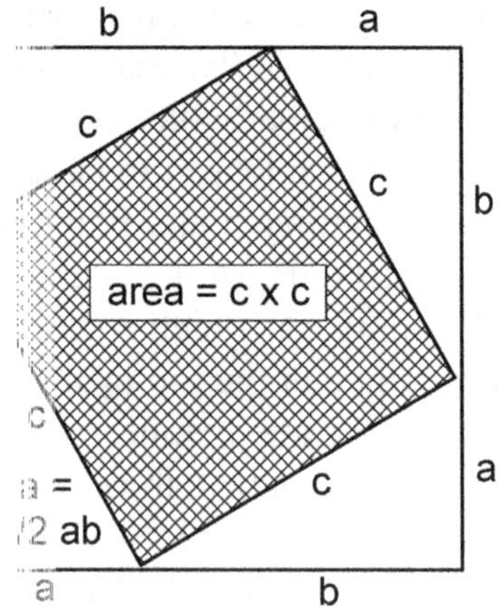

... Proof of Pythagorean theorem

Degrees an...

Our circle i... 360 ... wa... by the Babylonians. It could as well be divided int... *arbitra*... Th... *60* has the advantage, however, that it can be evenl... , 2, 3 ... 9 a...

We can als... circle ... s. T... erives from the intrinsic physical structure ... and h... art ... sider the structure of the circle in Fig. 7-4. The ratio of the *length* of the circumference to the diameter of the circle is a *constant* which we call π,

$$\frac{circumference}{diameter} = \pi$$

No matter what the size of the circle, the ratio of the circumference to diameter will *always* be the same.

But since the radius is equal to ½ the length of the diameter, we also have that

$$\frac{circumference}{2\ radii} = \pi$$

$$or \quad \frac{circumference}{radius} = 2\ \pi$$

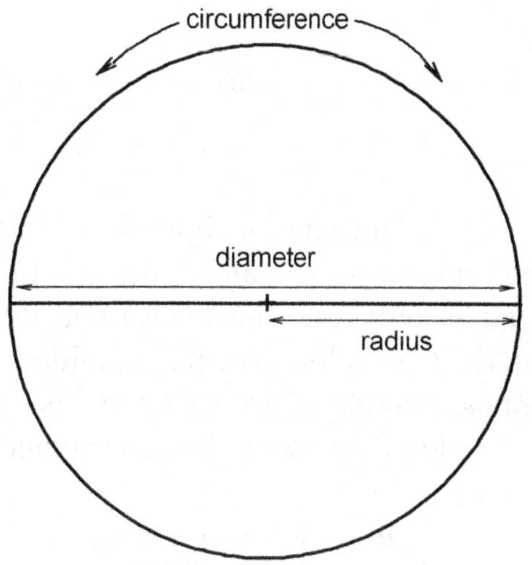

Fig. 7-4 Structure of circle

And since the circumference is also equal to 360 degrees (in angular measurement), we have

$$\frac{1/2\ circumference}{\pi} = radius; \quad we\ define\ \textbf{radian}\ to\ be\ \frac{180^o}{\pi\ (3.1416..)} = 57.296..^o$$

This lets us specify the degrees in a circle, or of an angle, either in degrees or **radians** where one radian equals ~57.3°. The unit is termed the radian because it is based on the *radius*. Most scientific calculators have a key that lets you enter the angle in either degrees or radians.

In some cases, it may be preferable to show the radians in terms of π, e.g., *360° = 2π radians, 180° = π radians, 90° = π/2 radians, 45° = π/4 radians, 30° = π/6, and so on.*

Calculation of pi, π

The constant **pi (or π)** is a fundamental constant widely used in mathematics. It is an *irrational* number with a value of *3.141592564...*, where the decimal numbers *never* go into a repeating sequence. In 2002, Japanese mathematicians used a supercomputer to calculate π to more than *1.24 trillion* decimal places, and found no evidence of a repeating sequence.

Archimedes (*287-212* BC) calculated π on the basis of the figure shown in Fig. 7-5. The circle is encompassed by an *external* six-side polygon, but the circle itself encompasses an *interior* six-sided polygon. There is also a radius (*radius-2* in Fig. 7-5) that is normal (i.e., at right angles) to the side of *both* the internal and external polygons. This yields two right angles. In the *interior* right angle, *radius-1* is the hypotenuse, and the "side opposite" equals ½ of the side of the interior polygon. Let θ be the *total* angle opposite the polygon side, and then sine (see below) of $\theta/2$ equals (1/2 side of interior polygon)/(radius-1), or

$$\sin\frac{\theta}{2} = \frac{1/2\ \textit{side of interior polygon}}{\textit{radius}} \qquad (\textit{since radius} - 1\ \textit{is just a radius})$$

$$\textit{and so}\ \ 2 \times \textit{radius} \times \sin\frac{\theta}{2} = \textit{side of interior polygon}$$

In the exterior right triangle, *radius-2* is the "side adjacent", and ½ the side of the *exterior* polygon is the "side opposite". Since tan $\theta/2$ (see below) is equal to (side opposite)/(side adjacent), tan $\theta/2$ is equal to (1/2 side of exterior polygon)/(radius). Thus,

$$\tan\frac{\theta}{2} = \frac{1/2\ \textit{side exteror polygon}}{\textit{radius}}$$

$$2 \times \textit{radius} \times \tan\frac{\theta}{2} = \textit{side of exterior polygon}$$

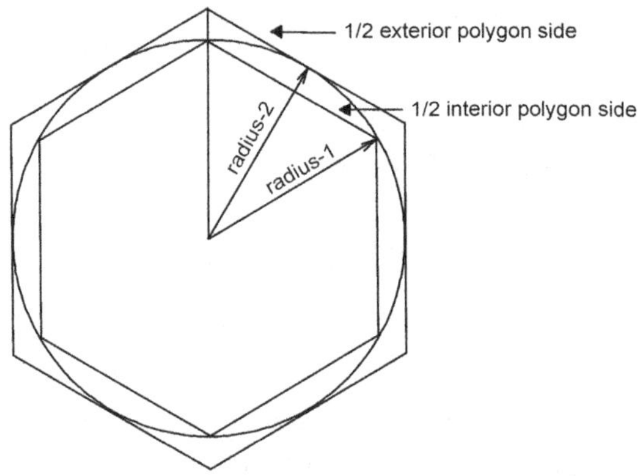

Fig. 7-5 Figure for calculation of pi

The perimeter of a polygon is equal to the number of sides × the length of the side. As the number of sides, *n*, becomes larger, the perimeter of the polygon approaches the *circumference* of the encased circle. And, if we set the diameter equal to *1*, the perimeter of the polygon tends toward π, since circumference/(diameter = 1) = π. We can thus calculate the perimeters for both polygons (and an approximation of π) according to

$$\textit{diameter} = 1;\ n = \textit{number of sides};\ p = \textit{perimeter};\ r = \textit{radius} = \frac{1}{2}\ \theta = \textit{angle}$$

internal polygon — *external polygon*

$$p = 2r\sin\frac{\theta}{2}\ n \approx \pi \qquad\qquad p = 2r\tan\frac{\theta}{2}\ n \approx \pi$$

Table 7-1 shows the calculation of π according to the equations above as the number of sides, *n*, becomes larger, and the internal angle, θ, becomes smaller.

Table 7-1

Approximation of the Value of π

Sides	Angle θ	Inter. Poly. Perimeter	Value of π	Exter. Poly. Perimeter
6	60°	3.0000...	3.14159...	3.4641...
24	15°	3.1326...	3.14159...	3.1596...
360	1°	3.1415...	3.14159...	3.14167...
720	0.5°	3.14158..	3.14159...	3.14161...

The area of a circle

The area of a circle is given by **area of circle = π r²**. Why should this be? One answer can be found using "magic scissors." These scissors can divide any area into an infinite number of parts, and when the parts are joined back together (in any arrangement) they yield *exactly* the same area. In Fig. 7-6, we cut a circle into four quadrants (four *90°* segments), and then rearrange these quadrants into a "twisty" figure. The area of this figure *has to be* just the same as the area of the circle. Next we cut the circle into twelve *30°* segments, and arrange these segments into a figure that, as we can see, approximates a rectangle.

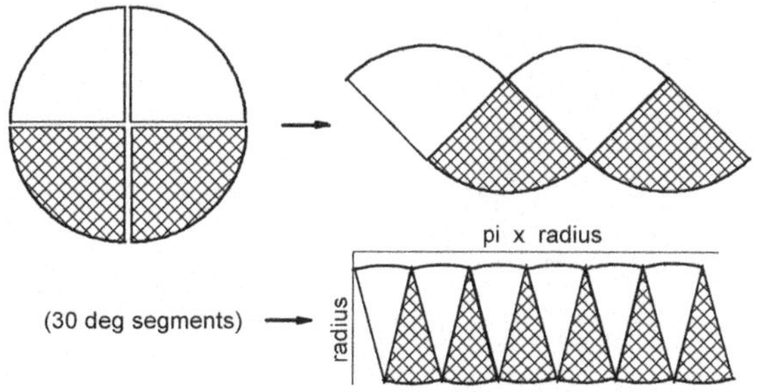

Fig. 7-6 Calculation of area of circle

Moreover, as the segments become smaller and smaller, the figure approaches a *limit* which is a rectangle where the short side equals **r** (since the leg of a segment is always equal to the radius), and where the long side equals **π x r** (since *π x r* is equal to ½ of the circumference of the circle). And since the area of the original circle must be equal to the area of this rectangle, we have **area of circle = π r²**

Trigonometric Functions

The common trigonometric functions are the **sine**, **cosine** and **tangent**. These functions are defined in accord with Fig. 7-7. In this figure, we rotate the radius in a *counterclockwise* manner, and generate the angle, φ. If we "drop a perpendicular" from the radius-circumference intersection to the horizontal diameter, we create a **right triangle**. The *basic functions* of angle φ are then defined as

$$\sin\varphi = [illegible] \frac{opposite}{hypotenuse}$$

$$\cos\varphi = [illegible] \frac{adjacent}{hypotenuse}$$

$$\tan\varphi = [illegible] \frac{opposite}{adjacent}$$

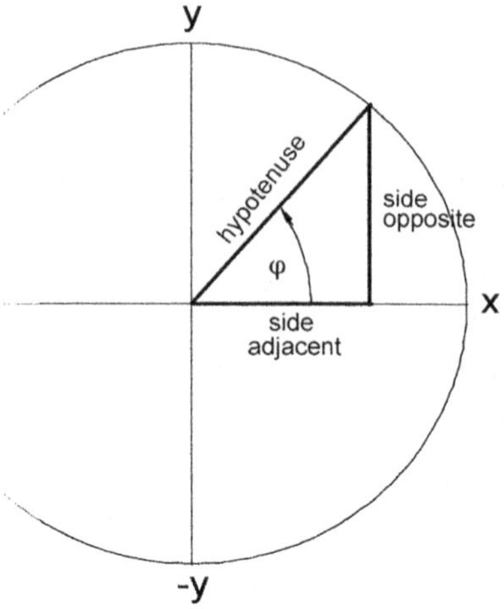

7 Right Angle Relationships

Present-day [illegible] provid[illegible] [illegible]s of the trigonometric f[illegible] [illegible] tou[illegible] [illegible]on. In earlier times, the[illegible] [illegible] to b[illegible] [illegible] in "trig" tables, or calc[illegible] [illegible] Taylo[illegible] [illegible] later) functions, e.g.

$$\sin x = [illegible] \frac{x^5}{5!} \ldots$$

We see from Fig. 7[illegible] [illegible] angle φ g[illegible] [illegible] 0 to [illegible] *sin φ* increases from *0 to 1*, but *cos φ decreases* fro[illegible] [illegible] sin[illegible] [illegible] of [illegible] ever be greater than *1*. On the other hand, the tan[illegible] [illegible] go f[illegible] [illegible] *inf*[illegible] [illegible]s from *0 to 90* degrees.

The trigono[illegible] [illegible]ions ar[illegible] [illegible]ed just to the *0 to 90* d[illegible] [illegible] of ang[illegible] [illegible] example, your calcu[illegible] [illegible]ow s[illegible] [illegible] to be *-0.500...*, and *co*[illegible] [illegible] to be [illegible] [illegible]We can understand this [illegible] [illegible]. Th[illegible] opposite" projects [illegible] [illegible] is for [illegible] *-0.500...*, and the "si[illegible] [illegible] proj[illegible] [illegible] axis for a length of [illegible] he ra[illegible] [illegible] to *1*, so the sine (side [illegible] [illegible]otenu[illegible] [illegible] is *-0.500.../1*, or *-0.50*[illegible] [illegible]ly, th[illegible] *210°* is *-0.866.../1*, [illegible] he t[illegible] *10* is *-0.500.../-0.866...* [illegible]

Notice that [illegible] the va[illegible] trigonometric funct[illegible] [illegible]nge [illegible] upon the quadrant, [illegible] [illegible]adran[illegible] numbered in a coun[illegible] [illegible] ma[illegible]

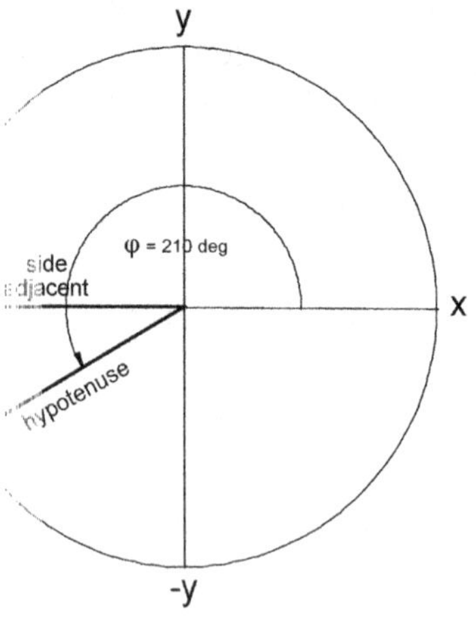

igonometric functions at 210 deg

Degrees	*Quadrant*	sin	cos	tan
0–90	1	+	+	+
90–180	2	+	–	–
180–270	3	–	–	+
270–360	4	–	+	–

Students are sometimes confused when they see the expression, "...*sin φ*" in an equation. They fail to realize that, although *φ* itself is an angle and has units of degrees or radians, the function, *sin φ*, is simply a decimal number of *1* or less. It has no units. Also, how to write the square of a *sin φ* function can be a problem. The term, $(\sin \varphi)^2$ is always correct, but lengthy. The accepted procedure is to write the square as $\sin^2 \varphi$.

Combining the right angle in Fig. 7-8 with the Pythagorean theorem gives a simple relationship between *sine* and *cosine* of angle φ. According to the Pythagorean theorem, the square of the hypothenuse is equal to the sum of the squares of the two sides, or $\mathbf{c^2 = a^2 + b^2}$. In Fig. 7-8, the hypothenuse is the radius which we define to have a length of *1*. The length of the "opposite side" is *sin φ (x 1)*, while the length of the "adjacent side" is *cos φ (x 1)*. Therefore, we have that

$$1^2 = \sin^2\varphi\ (1)^2 + \cos^2\varphi\ (1)^2, \quad or\ 1 = \sin^2\varphi + \cos^2\varphi$$

If we should happen to know *sin φ*, we can always find *cos* φ by

$$\cos \varphi = \pm \sqrt{1 - \sin^2 \varphi}$$

This is an important identity to remember in trigonometry. Whether *cos φ* is positive or negatives depends on the quadrant in which the angle *φ* lies.

Surveying and Optical Rangefinding

These are the main historical uses of right-triangle trigonometry. In surveying, we measure the height of a building, tree, mountain, etc., by using the properties of the right triangle. In Fig. 7-9, if we can measure the *horizontal distance* between the surveying transit and tree, and read the *angle of elevation, φ*, to the top of the tree, we can

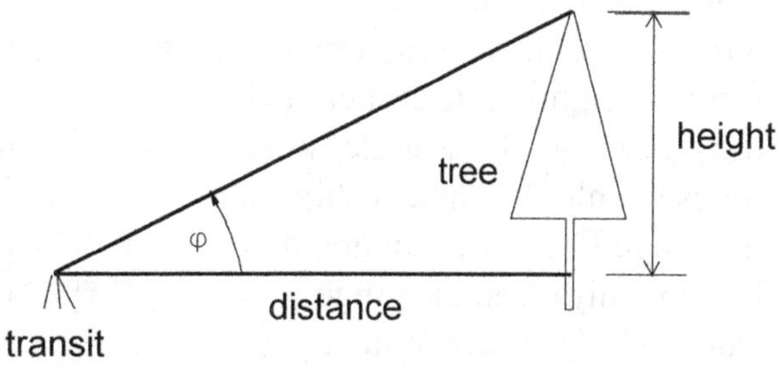

Fig. 7-9 Measuring height of tree

calculate the height of the tree by the *tangent function* of angle φ. The relevant equation is

$$\tan \varphi = \frac{side\ opposite}{side\ adjacent} = \frac{tree\ height}{distance}$$

If the distance between transit and tree is *150 feet*, and the angle of elevation is *33.15°*, then the height of the tree is

$$\tan 33.15^o = \frac{tree\ height}{150\ feet},\ or\ 0.653... \times (150\ feet) = 97.95\ feet = tree\ height$$

But, since the height of the horizontal line of sight from the transit to the tree is about *5 feet*, we must add *5 feet* to the indicated height of *97.95 feet*, giving the actual height of the tree above ground level to be *102.95 feet.*

Modern-day surveying uses instruments known as **theodolites** which, in their highest precision form, are capable of measuring angles to about *0.1* seconds (of **arc**). A degree is divided into *60* minutes, and a minute into *60* seconds, so there are *3600* arc-seconds in a degree. Measuring to *0.1* arc-seconds is equivalent to measuring to 0.1/3600 or 2.778×10^{-5} of a degree. The tangent of an angle is equal to the side opposite/side adjacent, and for an error in angle of 2.778×10^{-5}, and a side adjacent of *5280 ft*, the error in the side opposite would be

$$\tan 2.778\times10^{-5} = \frac{x}{5280\ ft},\ or\ x = (4.849\times10^{-7})\times5280 = 0.00256\ ft\ or\ 0.03\ \frac{in}{mi}$$

Surveyors use "first-order" to describe surveying data good to *1* part in *100,000*, or about *0.6* of an inch in one mile. The theodolites used in routine construction work can read angles to *5-20* arc-seconds which gives a potential error of *1.5 to 6* inches per mile.

The **optical rangefinder** on a battleship also uses a right triangle, but it is "turned on its side, and looked at from the other end." A schematic is shown Fig. 7-10. US battleships of the North Carolina class in World War II used a range finder with a base length of *26 ft. 6 in.*, or *8.0772 meters*. To find a range, the two telescopic images from the range finder, one being from the right-angle corner, and the other from the φ angle, were focused onto the subject ship (n.b., the range finder was mounted horizontally), and the range calculated from trigonometry. Assume the range in question to be *10,000 meters*. Then *tan φ* is

Fig. 7-10 Optical rangefinding

equal to side opposite over side adjacent, and we find φ by

$$\tan \varphi = \frac{10{,}000}{8.0772} = 1238.052791$$

$$\textit{then} \text{ arc } \textit{tangent of } 1238.502791 = \varphi = 89.9537_2$$

We can calculate the precision of the angle φ measurement that is necessary to find the range to within *100 meters* according to

$$\tan \varphi = \frac{10{,}100}{8.0772} = 1250.433318$$

$$\textit{then} \text{ arc } \textit{tangent of } 1250.433318 = \varphi = 89.9541_8$$

$$\textit{and, } \Delta\varphi = 89.9541_8 - 89.9537_2 = 0.00046 \textit{ degrees}$$

When this difference is expressed in **seconds of arc**, we see that it is equal to *1.7* seconds of arc,

$$\frac{x}{3600} = 0.00046 \textit{ degrees, or } x = 3600 \times 0.00046 = 1.7 \textit{ seconds of } \text{arc}$$

The accuracy of angle reading in battleship rangefinders must therefore be high, and approaches that of the highest level of theodolites (which are not subjected to naval battle conditions).

Terrain problems in surveying. The terrain often causes difficulties in surveying. One common problem is determining the elevation above sea level for sites within a hilly region (see Fig. 7-11). In these cases, one uses the theodolite as a **level** and measures the differences in height between the various sites on a **surveying rod** calibrated to *0.01 ft.*

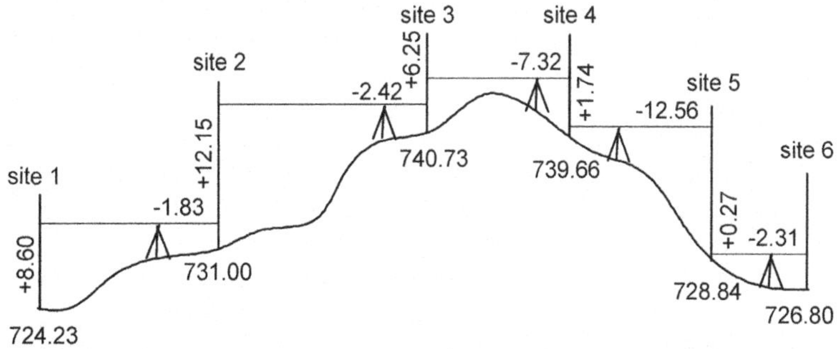

Fig. 7-11 Elevation of hilly terrain (in ft.)

The measurements, and calculations, are recorded in table form as shown here.

The abbrevi[illegible] FS [illegible] "bac[illegible] "forward shot", referring to the direction in which [illegible] was tak[illegible]

Site	[illegible] Elevation	[illegible]	FS
1	24.23		-1.83
2	31.00		-2.42
3	40.73		-7.32
4	39.66		-12.56
5	28.84		-2.31
6	25.80		
	-2.57 ft		-26.44 (= +2.57)

Using the ta[illegible] provides [illegible] the [illegible]rence in elevation between site *6* and site *1* is *2.57* [illegible] exac[illegible] the [illegible] between the sum of elevations

Fig[illegible] Measu[illegible] moun[illegible]ak height

added from the **BS** [illegible]s, a[illegible] of ele[illegible]*tracted* from the **FS** readings.

Another dif[illegible] suri[illegible] peaks is that the distance to the *perpendicular base* [illegible] ain [illegible] meas[illegible]. The situation is shown in Fig. 7-12. From a [illegible] of [illegible] angles of elevation, *Angle 1* and *Angle 2*, that are re[illegible] theod[illegible] straight line. The actual "up hill and down" dista[illegible] the [illegible] accurately measured, but we know (by the mean[illegible] Fig. 7-[illegible] se[illegible] *74 ft* higher than the *5000 ft* elevation of the first site. The measured angle of elevation between site *1* and site *2* is $1.245_5{}^o$. And therefore the distance between the two theodolites, as measured at the *5000 ft* elevation level, is given by

$$\tan(1.245_5^o) = \frac{574\,ft}{side\ adjacent}, \ or\ side\ adjacent = \frac{574\,ft}{0.02174} = 26{,}401\,ft = 5\,mi\ (26{,}400\,ft)$$

The distance between the bases of the theodolites at the *5000 ft* elevation level is thus 5 *mi* or *26,400 ft*. Also, the second theodolite is at an unknown distance, *x*, from the perpendicular base of the mountain.

The measured value for angle *1* in Fig. 7-12 is *7.706°*, while the measured value of angle *2* is *11.778°*.

First we must find *x*. Let us "slide back down" the line of sight of angle *2* to the *5000 ft* elevation. This does not change the angle *2* reading of *11.778°*. And we have then

$$\tan 11.778^o = \frac{574\ ft\ (side\ opposite)}{side\ adjacent}\quad or\quad side\ adjacent = \frac{574\ ft}{0.2085} = 2753\ ft$$

Referring back to Fig. 7-12, we see that

$$mtn\ ht = \tan_1\times(26{,}400 - 2753 + x) = \tan_2\times(x),\ or\ \ \tan_1\times(23{,}647)) = (\tan_2 - \tan_1)x$$

$$\tan(7.706^o) = 0.1353...,\ and\ \tan(11.778^o) = 0.2085...$$

$$thus,\ 0.1353(23{,}647) = (0.2085 - 0.1353)\,x,\ and\ \frac{3199\ ft}{0.0732} = 43{,}708\ ft = x$$

Once *x* is known, the mountain's height can be calculated according to

$$\tan(11.778^o)\ \frac{mtn\ ht}{x = 43{,}708},\ and\ \ mtn\ ht = 43{,}708 \times 0.2085 = 9{,}113\ ft$$

$$and,\ \tan(7.706^o)\ \frac{mtn\ ht}{43{,}708 + 26{,}400 - 2753}\ \ so\ \ mtn\ ht = 67{,}355\times 0.1353 = 9{,}113\ ft$$

The mountain's elevation above sea level thus comes out to be *5000 ft* + *9,113 ft* = *14,113 ft*. This is essentially the height of Pikes Peak which, at *14,115 ft*, can be seen for *130* miles from the eastern Colorado plains.

Other Types of Triangles

The right triangle is the most important in technology. But several other types of triangles are studied in trigonometry. These other types are shown in Fig. 7-13.

In an **equilateral triangle**, all three sides of the triangle are of equal length, and all three angles are 60°. In the **isosceles triangle**, two sides are of equal length, and the angles opposite these sides (Angle A) are equal. If all three sides of the triangle are of unequal length, the triangle is termed a **scalene triangle.** Scalene triangles are divided into the **acute triangle**, where all angles are smaller than 90°, and the **obtuse triangle**, where *one* angle is larger than 90°. Triangles that are not right triangles are called **oblique triangles**.

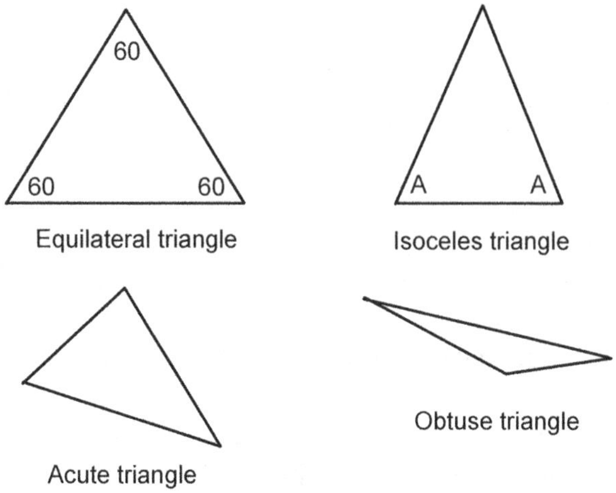

Fig. 7-13 Different triangles

The sum of the three angles within a triangle *must* be equal to 180°. One way to understand this is to look at the obtuse triangle in Fig. 7-14. The radius has been rotated counterclockwise to nearly 180°, and the interception points of the radii with the circumference have been connected with a long line to form an obtuse triangle. As the rotation continues, Angles A and C become infinitesimal, and Angle B approaches 180°. And the sum of the angles, $0^o + 180^o + 0^o$, becomes equal to 180°.

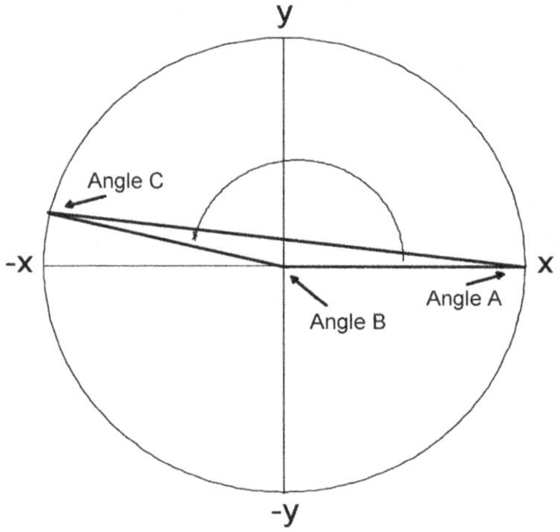

Fig. 7-14 Generation of obtuse triangle

Area of a triangle

The area of *any* triangle can be found by the equation, $A = \frac{1}{2}bh$, where A is the area of the triangle, b is the *base* of the triangle, and h is the *height* of the triangle. The figure of a triangle is commonly presented with the base being horizontal, and the height perpendicular to the base, as with the right and isosceles triangles shown in Fig. 7-15.

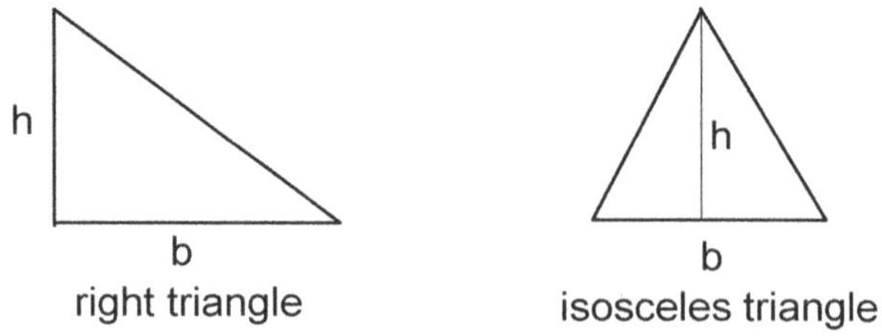

Fig. 7-15 Triangles for calculating area

Notice that if we multiply the base by the height, we obtain a rectangle whose area is just *twice* that of the triangle. The area of these types of triangle is therefore given by

$$area = \frac{1}{2}\ (base \times heighth) = \frac{1}{2}\ bh$$

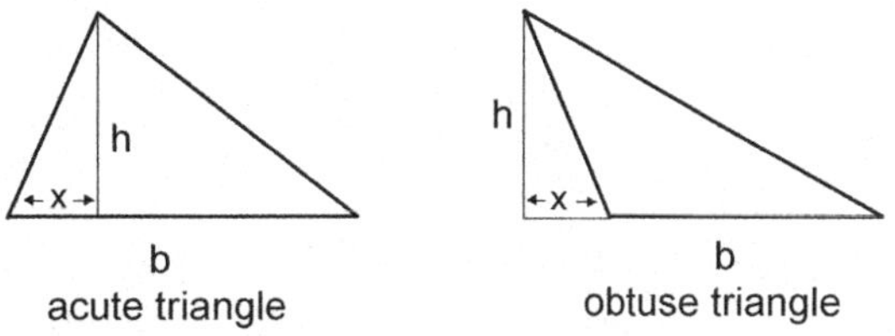

Fig. 7-16 Areas of acute and obtuse triangles

This same equation holds for acute and obtuse triangles also. However, the "proof" is somewhat more difficult to envision. Let us look at Fig. 7-16.

With the acute triangle, we can flip over the small right triangle to the left of the height line, and add it to the original right triangle. The area of the resultant rectangle is $(x)h$. Doing the same thing with the larger right triangle to the *right* of the height line produces a rectangle of area $(b - x)h$. The sum of these two areas is $(x)h + (b-x)h = bh$, and ½ of the area of this "added-up" rectangle is the area of the acute triangle, i.e., $\frac{1}{2}\, bh$.

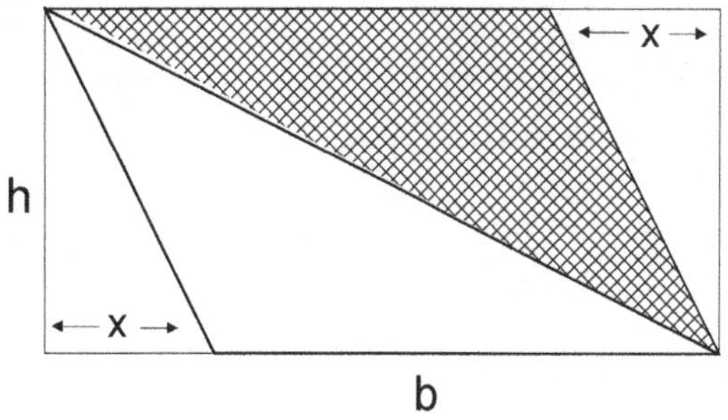

Fig. 7-17 Area of obtuse triangle

With the obtuse triangle, we can form a rectangle by multiplying $b + x$ by h. Its area is equal to $(b + x)h$. This area is *larger* than the obtuse triangle, however, by the areas of the right triangles at the left, and at the right, of the rectangle, as shown in Fig. 7-17. The area that remains is the original obtuse triangle plus its flipped-over *double*, as indicated by the hatched area. The area of the rectangle that represents the doubled obtuse triangle is therefore $(b + x)h - 2(1/2\ xh) = bh$. And one half of that area is the area of the obtuse triangle, or

$$area\ of\ obtuse\ triangle = \frac{1}{2}\ bh$$

The formula which says that the area of a triangle is equal to ½ *base* × *height* is thus valid for *all* triangles.

Law of Sines

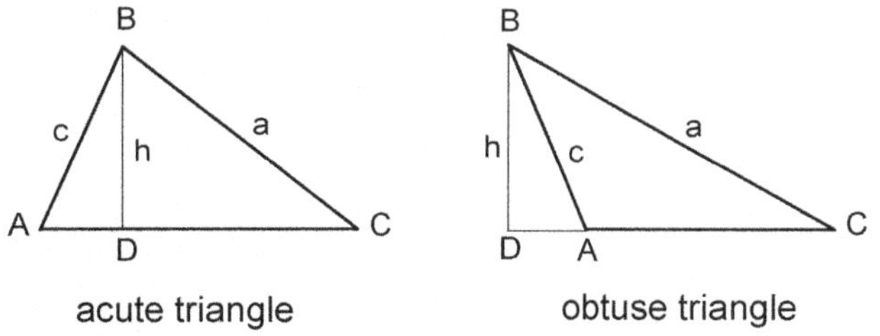

Fig. 7-18 Law of sines figures

In addition to the area of a triangle, we often wish to know its proportions. We can do this with the **law of sines**. Looking at the triangles ABD and DBC in the acute triangle in Fig. 7-18 (and noting that h is the height and not the hypotenuse) , we see that *[illegible]DB*, [illegible] $\frac{h}{c}$ [illegible] sin A

[illegible]BDC, [illegible] $\frac{h}{a}$ [illegible] sin C

[illegible]herefo[illegible] [illegible] A [illegible]

and [illegible] [illegible]ugh [illegible] [illegible] $\frac{a}{\sin A} = \frac{c}{\sin C}$

If we drew a perpe[illegible] [illegible] angl[illegible] [illegible] (ra[illegible] [illegible]om angle B to side b as in Fig. 7-18), we could sh[illegible] [illegible] [illegible]r arg[illegible] [illegible] *b/*[illegible] *C*, and have the complete law of sines, i.e.,

[illegible] = $\frac{b}{\sin}$ [illegible] $\frac{c}{[illegible] C}$ [illegible] *[illegible]f sines*

This same d[illegible] [illegible]ks f[illegible] [illegible]e tri[illegible] [illegible]g. 7-18. Students are sometimes bothered by the fac[illegible] [illegible]ght, *h*[illegible] [illegible] the[illegible] [illegible]C. However, the definitions still hold. The sine [illegible] [illegible] equal [illegible] [illegible]oppo[illegible] [illegible]enuse, or *h/a*. And although the angle A is large[illegible] [illegible] sine i[illegible] [illegible]al to [illegible] [illegible]posite/hypothenuse or *h/c*. Recall that when A [illegible] [illegible] *90°*, i[illegible] [illegible]ust e[illegible] [illegible] sine of (*180°* - A). For example, the sine o[illegible] [illegible]*66*, bu[illegible] [illegible]f (18[illegible] [illegible] 60°) is also *0.866*.

We have th[illegible] [illegible] [illegible] obtu[illegible] [illegible], jus[illegible] [illegible]acute triangle, that

[illegible]CAB, [illegible] $\frac{h}{a}$ [illegible] sin C

[illegible]BD, [illegible] $\frac{h}{c}$ [illegible] sin A

[illegible]herefo[illegible] [illegible] A [illegible]

and as [illegible] [illegible]ute [illegible] [illegible] h[illegible] [illegible] $= \frac{c}{\sin C} = \frac{b}{\sin B}$

Notice that [illegible] [illegible]f si[illegible] [illegible] de[illegible] [illegible] one term if you know the other three terms. If for [illegible] [illegible]e A [illegible] [illegible]C i[illegible] [illegible]= *100 meters*, then

$$\frac{a}{\sin 35^o} = \frac{100\ meters}{\sin 47^o};\quad a = \frac{0.574}{0.731}(100\ meters) = 78.5\ meters$$

$$and\ \frac{b}{\sin\ (180-47-35=)\ 98^o} = \frac{100}{\sin 47^o};\quad b = \frac{0.990}{0.731}(100\ meters) = 135.4\ meters$$

then for ΔABC, $A = 35^o$, $B = 98^o$, $C = 47^o$, *and* $a = 78.5\ m$, $b = 135.4\ m$, $c = 100\ m$

All of the dimensions of the triangle can therefore be found using the law of sines. Note that side *a* is across from angle *A*, side *b* is across from angle B, and side *c* is across from angle C. Also, the *longest* side is always across from the *greatest* angle.

A triangle can not be constructed from just any three lengths of line. Lines of *10 meters*, *4 meters* and *5 meters* in length can not be combined, for example, to make a triangle. The sum of the lengths of the two shortest lines must be *more* than the length of the longest line. We can see this by considering Fig. 7-19.

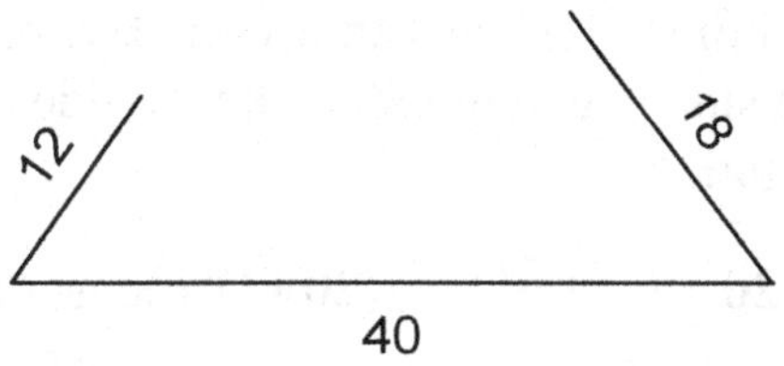

Fig. 7-19 Impossible sides of triangle

Law of Cosines

The law of cosines is another means for calculating the dimensions of a triangle. It is more restrictive however, since it requires that you know the lengths of two sides of the triangle and the angle *between* these two sides. The law of cosines is derived as follows, using Fig. 7-20.

Any triangle can be divided into two right triangles by drawing a perpendicular from *at least* one of its angles to the opposite side of the triangle. In Fig. 7-20, the sine of C is *h/a*, while the cosine of C is s_1/a. Thus

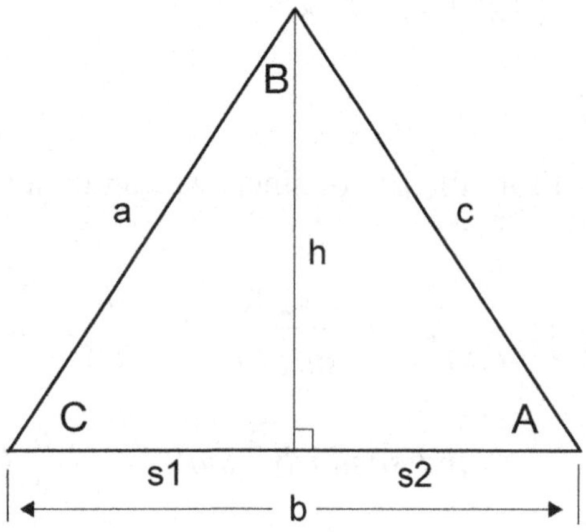

Fig. 7-20 Figure for cosine law

$$h = a \sin C;\quad and\quad s_2 = b - s_1 = b - a \cos C$$

Then looking at the right triangle on the right side of Fig. 7-20, we have

$$c^2 = h^2 + s_2^2$$

$$c^2 = (a \sin C)^2 + (b - a \cos C)^2$$

$$c^2 = a^2\sin^2 C + b^2 - 2abcosC + a^2\cos^2 C$$

$$and\ then\quad c^2 = a^2(\sin^2 C + \cos^2 C) + b^2 - 2abcosC$$

$$so\quad c^2 = a^2 + b^2 - 2abcosC$$

Note that the last line of the derivation makes use of the fact that $sin^2C + cos^2C = 1$.

In Fig. 7-21, we can use the law of cosines to calculate side *c*, given that we know sides *a* and *b*, as well as angle *C*.

$$c^2 = 20^2 + 31.6^2 - 2(20)(31.6)\cos(110^o)$$

$$= 400 + 998.6 - 1264(-0.342) = 1830.9$$

$$c = \sqrt{1830.9} = 42.8$$

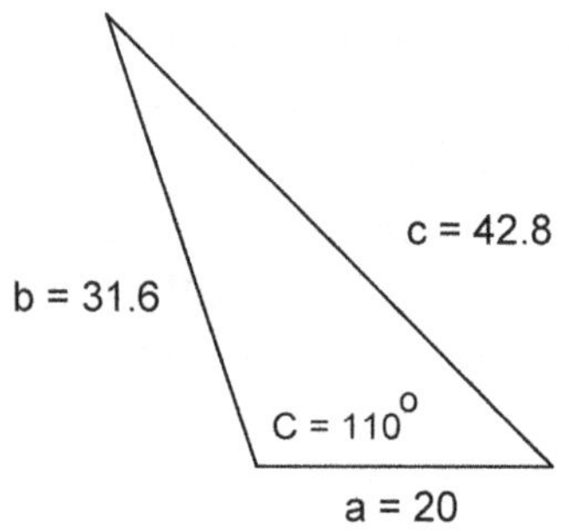

Fig. 7-21 Law of cosines calculation

From the law of sines, we can then calculate the remaining angles

$$\frac{c}{sinC} = \frac{a}{sinA},\quad or\quad \frac{42.8}{sin110^o} = \frac{20}{sinA};\quad sinA = \frac{20}{42.8}(0.940) = arcsinA = 0.439;\quad A = 26^o$$

$$and\ similarly\quad sinB = \frac{31.6}{42.8}(0.940) = \text{arcsin}B = 0.694;\quad B = 44^o$$

$$Total\ angles\ in\ triangle = 110^o + 26^o + 44^o = 180^o$$

Trigonometric Identities

In addition to the "basic" trigonometric functions of sine, cosine and tangent, mathematicians define and use the *reciprocal* functions of **cosecant (csc), secant (sec), and cotangent (ctn)**. All of these functions can be derived from the fundamental properties of an angle, as depicted in Fig. 7-7. We can summarize the functions as

Summary of Reciprocal Functions

Function	Reciprocal function
sin A = y/r	csc A = 1/sin A = r/y
cos A = x/r	sec A = 1/cos A = r/x
tan A = y/x	ctn A = 1/tan A = x/y

In this summary, note that (as in Fig. 7-7) the hypotenuse, *r*, is equal to the radius, and that *y* is the projection of the "side opposite" onto the *y* axis, while *x* is the projection of the "side adjacent" onto the *x* axis.

From these six trigonometric functions, a large number of trigonometric **identities** can be deduced. These identities are set in the form of *equalities* (meaning that something is equal to something else), but they are meant to record a specific trigonometric relationship, and not to be "solved" themselves. As an example, we see from Fig. 7-7 that $x^2 + y^2 = r^2$, and setting the angle φ to be angle A, we can show this relationship,

$$x^2 + y^2 = r^2, \textit{ and dividing by } x^2 \textit{ gives, } 1 + \frac{y^2}{x^2} = \frac{r^2}{x^2}$$

$$\textit{or from reciprocal functions,} \quad 1 + \tan^2 A = \sec^2 A$$

The **addition rule** is another form of trigonometric identity. This rule determines the value of the function *sin(A + B)* where *A* and *B* are two *known* angles. The solution is based on graphics where the two angles are added *graphically*, as in Fig. 7-22.

In Fig. 7-22, we see that sin (A + B) is equal to $(y_1 + y_3) / h_2$. We must determine y_3. We see also that $y_3 = y_2 \cos C$. Since they are angles in a right triangle, angles A and D must equal 90°. And angles C and E are also in a right triangle, so angles C and E must also equal 90°. Finally, since they are on the "opposite" sides of two intersecting straight lines, angle D must equal angle E.

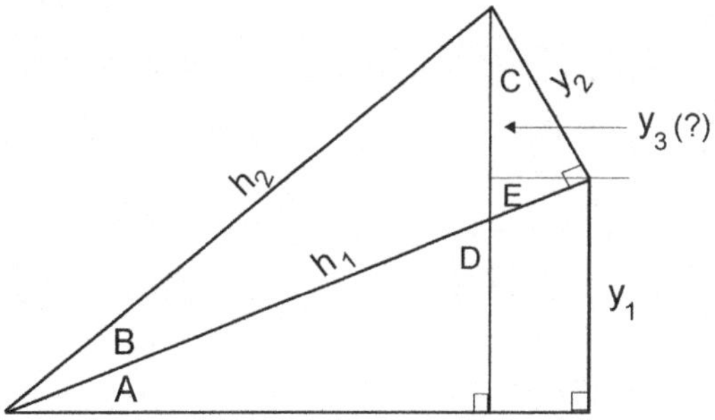

Fig. 7-22 Determination of sin (A + B)

We have therefore that

$$\textit{angle } A + \textit{angle } D = 90^o = \textit{angle } C + \textit{angle } E$$

$$\textit{and since angle } D = \textit{angle } E; \quad \textit{then angle } A = \textit{angle } C$$

which then leads to

s 1 (A + ... sin ... s A

From Fig. 7-22, we ... t at

... = h ... and ... B

When we make the ... ns, w... lly th

sin (A + B) ... sin ... B ... A cos B + sin B cos A

To test this identify ... e tha ... nd B ... ing our calculator, we then find that

s 1 49° ... (calc ... te)

... r 37° ... + ... 37°) =

...)(0.9 ... 07 ... 0.7546

There are m ... gono ... ions ... ng cofunctions, quotient relations, and suppl ... gle ide ... ese ... mulas for the difference between two angle ... he d ... of an angle (2A), f ... g of a ... (A/2), or for the sq ... angle ... These auxiliary fun ... all be ... worked out by mat ... sing essentially Fig. 7-7 ... is defi ... of sine, cosine and ...

Polar Coordina...

The positio ... r two dimensional space ... d by rectangular (x,y) co ... by pol **coordinates**. In x, ... s in Fi ... the origin is (0,0), ... t po ...

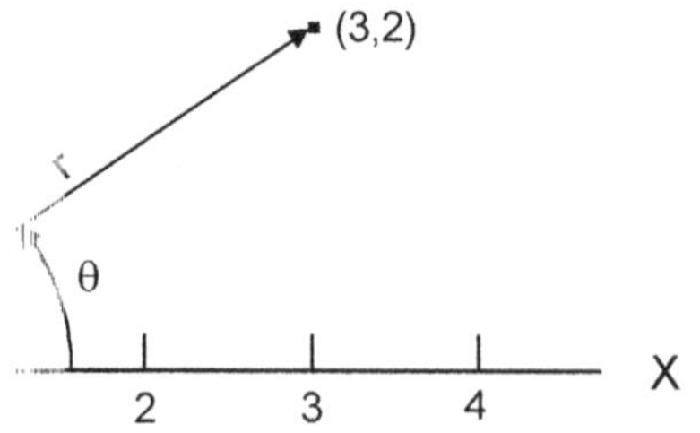

...olar coordinates

$(x,y) = (3,2)$. In polar coordinates, the distance from the origin to the subject point is shown by a *direct line*, r, which is rotated from the origin line $(y = 0)$ by θ degrees. The value of θ is given by the arctan of $2/3$, with θ therefore being 33.68^o. The length of r is $(3^2 + 2^2)^{0.5}$ or $13^{0.5}$, which equals *3.606*.

The relationship between rectangular and polar coordinates is given by

Conversion to polar coordinates (r, θ)

$$r = \sqrt{x^2 + y^2}$$

$$\theta = \arctan \frac{y}{x}$$

Conversion to rectangular coordinates (x, y)

$$x = r \cos \theta$$

$$y = r \sin \theta$$

Distance between two points. With rectangular coordinates, we can find the distance between two points from their indices. The distance between Point 1 (2,3) and Point 2 (5,6) is equal to the square root of $(x_2 - x_1)^2 + (y_2 - y_1)^2$ or $(5 - 2)^2 + (6 - 3)^2$; thus, the distance between the two points is equal to $(18)^{0.5}$ or ***4.243***. This is just an application of the Pythagorean theorem.

To find the distance between the same two points when polar coordinates are used, we see that, when the rectangular indices are converted to polar coordinates,

Point $1(x = 2, y = 3)$

$$\theta_1 = \arctan \frac{3}{2} = 56.310^o; \quad r_1 = \frac{x}{\cos \theta} = \frac{2}{0.555} = 3.606$$

Point $2(x = 5, y = 6)$

$$\theta_2 = \arctan \frac{6}{5} = 50.194^o; \quad r_2 = \frac{x}{\cos \theta} = \frac{5}{0.640} = 7.813$$

The distance between Points 1 and 2 is a straight line, and if it were drawn, it would produce a *triangle* in which we know the lengths of two sides *(3.606 and 7.813)* and the angle $(56.310^o - 50.194^o = 6.116^o)$ between the two sides. We can therefore use the **law of cosines** and find that

$$c^2 = a^2 + b^2 - 2ab\cos\theta, \textit{ where } \theta = 6.116^o$$

$$= (3.606)^2 + (7.813)^2 - 2(3.606)(7.813)(0.994)$$

$$c = \sqrt{18.03} = \mathbf{4.247}$$

The distance between points is just the same (within errors of calculation) as found with rectangular coordinates. This is expected since *only* the coordinates were changed. Depending upon the particular calculation, one may find either rectangular or polar coordinates to be the most convenient.

Complex Numbers

Complex numbers are a mixture of real and imaginary numbers. Recall that the imaginary number is i, which is equal to $\sqrt{-1}$. Complex numbers are written as

$$complex\ number = a + bi$$

$$where\ a,b\ are\ \underline{real}\ numbers$$

Complex numbers are customarily plotted as 2-dimensional plots using either rectangular or polar coordinates. The numbers are essentially *split up*, with the *x axis* representing the real number system, and the *y axis* representing the *imaginary* number system. With the rectangular system, we have

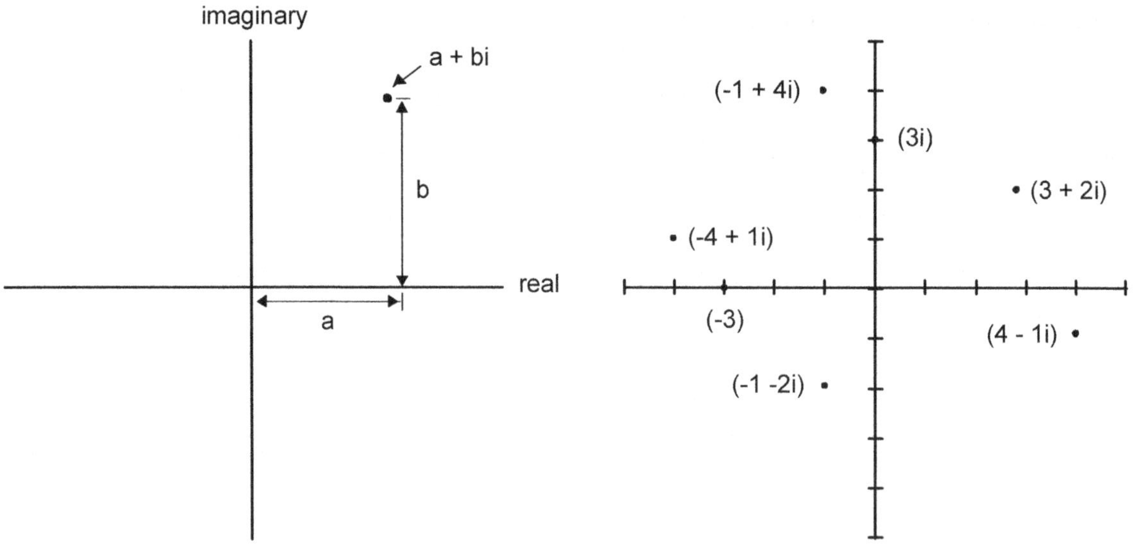

Fig. 7-24 Complex numbers using rectangular coordinates

Notice that the indices of the points are written as x units of *real* numbers plus (or minus) y units of *imaginary* numbers. This emphasizes that the two axes deal with different types of numbers, and that these numbers must be handled differently.

If we look at arithmetic operations with complex numbers, we find the following behaviors:

Addition-Subtraction. Complex numbers expressed in rectangular coordinates add and subtract in a straightforward manner, as for example,

$$\textit{add}\quad \begin{array}{r} 5 - 3i \\ +(4 - 2i) \\ \hline 9 - 5i \end{array} \qquad \textit{subtract}\quad \begin{array}{r} 5 - 3i \\ -(4 - 2i) \\ \hline 1 - i \end{array}$$

Addition and subtraction can also be done *graphically* with points that are plotted on the complex plane. It may be simpler however to calculate the coordinates of the new point algebraically, and then plot the new point.

Multiplication-Division. Complex numbers given in rectangular coordinates also multiply algebraically with no difficulty:

$$\begin{array}{r} 5 - 3i \\ \times\ 4 - 2i \\ \hline 20 - 12i \qquad\quad \\ - 10i + 6\,i^2 \\ \hline 20 - 22i + 6\,i^2 \end{array}$$

$$\textit{or since } i^2 = -1, \textit{ the answer equals } 14 - 22i$$

But, because we converted i^2 to *-1*, we can *not* simply divide *14 - 22i* by *4 - 2i* to obtain the original multiplicand, *5 - 3i*. However, if we consider the multiplicand simply as x - y*i*, we can see by an analysis of the multiplication process that

$$\textit{coefficient of i term} = 4(-y) + (-2)x = -22, \textit{ or } -4y - 2x = -22$$

$$\textit{and coefficient of real term} = 4(x) + (-2)(-y)(i^2 = -1) = 14$$

$$\begin{array}{rr} \textit{rearranging and} \times 2 & -4x - 8y = -44 \\ & 4x - 2y = 14 \\ \hline & -10y = -30 \end{array}$$

$$\textit{so } y = 3 \textit{ and } x = 5, \textit{ and the multiplicand} = 5 - 3i$$

An alternat[illegible] l divi[illegible] of the divisor. If the divisor is $3 + 2i$, its conjugat[illegible] ltip[illegible] gether eliminates the i term, e.g. $(3 + 2i)(3 - 2i) = 9$ [illegible] $= 13$ [illegible] on of this type, consider

$$\frac{(5 + \ldots}{(3 + \ldots} \ldots \frac{\ldots + 5i)(\ldots}{\ldots + 2i)(\ldots} \ldots (15 \ldots i^2) = \frac{25}{13} + \frac{5}{13}\,i$$

Complex Numb[illegible] Polar [illegible]dinates

Multiplicati[illegible] ns of **exponents** are simpler with polar coordina[illegible] e elegant mathematics involving complex [illegible] oordinates.

Complex n[illegible] ord with Fig. 7-25. The x axis represents the *real* [illegible] axis the *imaginary* component. The location of any [illegible] en by [illegible] distance, r, from th[illegible] of th[illegible] plot, and the angle, [illegible] he r line and the positiv[illegible] horizontal *real* axis, [illegible] esen[illegible] a rotation in the co[illegible] e direction.

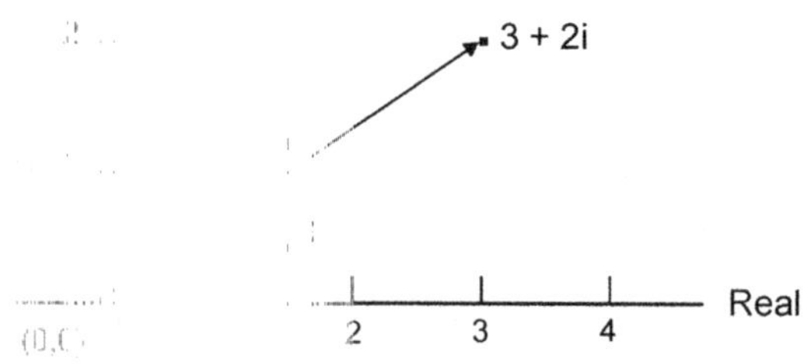

[illegible] numbers in polar coordinates

The fact tha[illegible] represents *units* of [illegible], whil[illegible] the y axis represent[illegible] gina[illegible] numbers must be ta[illegible] unt. However, the abso[illegible] is st[illegible] just equal to $\sqrt{(x^2 + \ldots}$ [illegible] cas[illegible] $\sqrt{(3^2 + 2^2)} = \sqrt{13} =$ [illegible] tangent of θ is side [illegible] adjacent, or $2/3 =$ [illegible] is th[illegible] θ is equal to $33.69°$. Using these values, we se[illegible] tion [illegible] call z, can be found according to

$$x = r(\cos\theta) \ldots 32) = \ldots \quad \ldots = 3.605(0.555) = 2.00\ldots$$

therefore [illegible] *of* [illegible] *(iy units) is given by*

$$\ldots \cos\theta \ldots \text{(in } \ldots \text{dinates)}$$

We wish also to de[illegible] on of [illegible] tes. To do this, we can use **Euler's formula**,

$$(\textit{exponential coordinates})\ e^{i\theta} = \cos(\theta) + i\ \sin(\theta)\ (\textit{polar coordinates})$$

There are several derivations of Euler's formula, but the one that uses a Taylor series about *0* (see Chap. 8 on series calculations) is probably easiest for students to follow. Here are the Taylor series representations of three important terms in mathematics:

$$e^x = 1 + \frac{x}{1!} + \frac{x^2}{2!} + \frac{x^3}{3!} + \frac{x^4}{4!} + \ldots$$

$$\sin x = \frac{x}{1!} - \frac{x^3}{3!} + \frac{x^5}{5!} - \frac{x^7}{7!} + \ldots$$

$$\cos x = 1 - \frac{x_2}{2!} + \frac{x^4}{4!} - \frac{x^6}{6!} + \ldots$$

The first equation was used by Euler to calculate the value of *e* (pg. 38) to eighteen decimal places. If we let $x = 1$, then we have

$$e^1 = 1 + \frac{1}{1!} + \frac{1^2}{2!} + \frac{1^3}{3!} + \frac{1^4}{4!} + \ldots\ (= 2.7083\ldots)$$

As given by our calculator, the value of *e* is *2.718281828...* to nine places. Euler evidently used more terms to calculate his highly accurate value of *e*.

But let us go beyond this, and set e^x to be e^{ix}. And we note that $i^1 = i$, $i^2 = -1$, $i^3 = -i$, $i^4 = 1$, $i^5 = i$, $i^6 = -1$, $i^7 = -i$, *and so on*. This gives

$$e^{ix} = 1 + \frac{ix}{1!} + \frac{i^2x^2}{2!} + \frac{i^3x^3}{3!} + \frac{i^4x^4}{4!} + \ldots$$

$$= 1 + \frac{ix}{1!} - \frac{x^2}{2!} - \frac{ix^3}{3!} + \frac{x^4}{4!} + \frac{ix^5}{5!} + \ldots$$

Next let us multiply the series expression for *sin x* by *i*, which yields

$$i\ \sin x = \frac{ix}{1!} - \frac{ix^3}{3!} + \frac{ix^5}{5!} + \ldots$$

And if we add the series expression for *cos x* to this series, we obtain a series that is equal to the sum of *cos x + i sin x*, i.e.,

$$\cos x + i \sin x = 1 + \frac{ix}{1!} - \frac{x^2}{2!} - \frac{ix^3}{3!} + \frac{x^4}{4!} + \frac{ix^5}{5!} + \ldots$$

Comparison of this last series with the series expression for e^{ix} shows that they are just the same; therefore

$$e^{ix} = \cos x + i \sin x$$

In the calculations above, we have interchanged the variables θ and x, but both in fact refer to a measurement of the angles involved in terms of *radians*. If we let x equal π in radians (or 180° when the angles are measured in degrees), we then have that

$$e^{i\pi} = \cos \pi \; (= -1) + i \sin \pi \; (= 0)$$

$$\textit{and so} \quad e^{i\pi} + 1 = 0 \qquad (\textit{Euler's Identity})$$

Euler's Identity is just one special case of Euler's formula, but it is famous in mathematics, and is often called the "most beautiful" of mathematical theorems.

This illustrates once again the *power* of the equal sign. We find an equation for e^{ix} by quite unexpected (for non-mathematicians) simple algebraic manipulations of the Taylor series for e^x, $\sin \theta$ and $\cos \theta$. Moreover, this equation relates the polar coordinates to the rectangular coordinates of the complex number system, i.e.,

$$e^{i\theta} = \cos \theta + i \sin \theta \quad (\textit{substituting } \theta \textit{ for } x)$$

And since the location of z in the real number-imaginary number plane is given by

$$z = x \; (= r \times \cos \theta) + i\, y \; (= r \times \sin \theta) = r\,(\cos \theta + i \sin \theta)$$

we have that

$$z = r(e^{i\theta})$$

These relationships can also be manipulated to produce the classic, but exotic-looking identities,

$$\cos\theta = \frac{e^{i\theta} + e^{-i\theta}}{2} \quad \textit{and} \quad \sin\theta = \frac{e^{i\theta} - e^{-i\theta}}{2i}$$

We begin by taking the *conjugate* of Euler's formula, where the conjugate is produced by replacing each i term with a $-i$ term. We then add (or subtract) the conjugate with the original Euler's formula, and solve for *cos θ* and *sin θ.*

$$(\textit{Euler's formula}) \quad e^{i\theta} = \cos\theta + i\sin\theta$$

$$(\textit{conjugate}) \quad e^{-i\theta} = \cos\theta - i\sin\theta$$

$$\textit{Euler's formula} + \textit{conjugate} = e^{i\theta} + e^{-i\theta} = 2\cos\theta; \quad \cos\theta = \frac{e^{i\theta} + e^{-i\theta}}{2}$$

$$\textit{Euler's formula} - \textit{conjugate} = e^{i\theta} - e^{-i\theta} = 2i\sin\theta; \quad \sin\theta = \frac{e^{i\theta} - e^{-i\theta}}{2i}$$

Example calculations. We can represent the position of any point (or value of z) on the real-imaginary plane in terms of three different coordinate systems,

Rectangular	*Polar*	*Exponential*
$x + iy$	$r(\cos\theta + i\sin\theta)$	$re^{i\theta}$

However, the mathematical relationships between the different points can not be changed by the coordinate system. We should find the same answer for any complex number operation regardless of whether it is given in rectangular, polar or exponential coordinate terms.

To test this idea, we will look at multiplication. It is difficult to conceive of how a *point* in the real-imaginary plane could be multiplied by another point. Would we end up with "points squared"? But we can multiply the *value* of each point, and find the value of a third point, which we can plot. Let point *1* equal $2 + i$, and point *2* equal $1 + 3i$, then the multiplication of point *1* by point *2* gives point *3*, which has a value of $-1 + 7i$:

$$\begin{aligned} \textit{Point } 1 &= 2 + i \\ \times\ \textit{Point } 2 &= \underline{1 + 3i} \\ \textit{Point } 3 &= 2 + i + 6i + 3i^2 \\ &= -1 + 7i \end{aligned}$$

These three points ... rect... coordinates in Fig. ... insp... might not suggest ... sults fr... multiplication of p... int 2, ... algebraic multiplic... that i...

Next let us ... ne p... polar coordinates. ... we ... a portion of Fig. 7-... essed ... location of point *1* ... polar ... From Fig. 7-27, we ...

$$r_1 = \sqrt{2^2 \ldots} = 2.2\ldots$$

$$\theta_1 = \text{arcta}\ldots \ldots 7^o$$

$$r_2 = \sqrt{3^2 \ldots} = 3.\ldots$$

$$\theta_2 = \text{arcta}\ldots \ldots 6^o$$

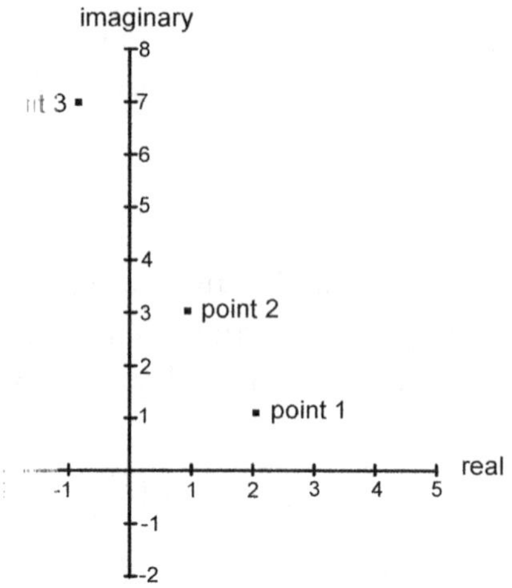

...tangular multiplication

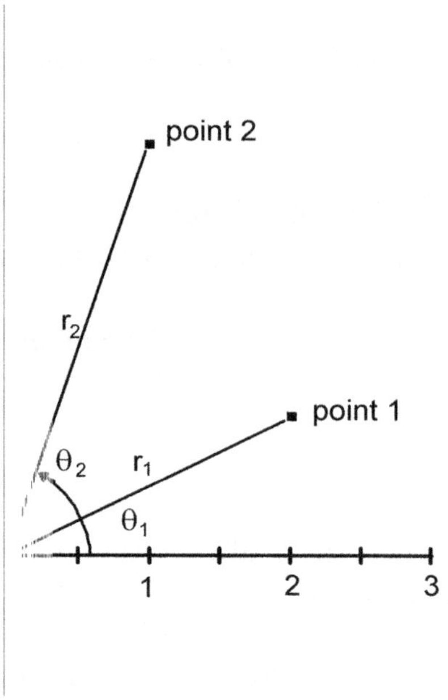

Polar coordinates

If we multiply the p... ... ate va... points *1* and *2*, we h...

$$r_1(\cos\theta_1 + i\sin\ldots \ldots \theta_2 + i\ldots$$

which ... *to*

$$r_1r_2[\cos\theta_1 \ldots \sin\theta_2 \ldots$$
$$i(\sin \ldots \theta_2\cos\ldots$$

Looking at the last ... that i... two trigonometric i... ...nely,

$$\cos\theta_1\ldots \ldots\sin\theta_2 \ldots + \theta_2 \ldots$$ (...*ig addition rules*)

and ... + sin... sin... (*from pg.* 106)

Therefore, to multiply two complex numbers (which are values of points on the real-imaginary plane), we simply use the relationship,

$$\textit{complex number product} = r_1 r_2[\cos(\theta_1 + \theta_2) + i \sin(\theta_1 + \theta_2)]$$

If we apply this equation, using the data from Fig. 7-27, we find that

$$\textit{complex number product} = (2.236)(3.162)[\cos(26.57 + 71.56 = 98.13^o) + i \sin(98.13^o)]$$

$$= -0.999_9 + i\,6.998_6 \approx -1 + 7\,i$$

Thus we have the same value for point *3* as we found with rectangular coordinates. Moreover, the arctan of *-1/7* is equal to *-8.13°*. From Fig. 7-27, we see that adding *8.13°* to the *90°* of the first quadrant gives *90° + 8.13° = 98.13°*. And so, $\theta_1 + \theta_2 = \theta_3$, as expected.

For the process of multiplication using exponential coordinates, we have that

$$z_1 \times z_2 = r_1 e^{i\theta_1} \times r_2 e^{i\theta_2} = r_1 r_2 e^{i(\theta_1 + \theta_2)} = r_1 r_2 e^{i\theta_3} = z_3$$

That is, if we multiply the value of point *1* by the value of point *2*, we obtain the value of point *3*. We know the values of r_1, r_2 and θ_3 from above, but we can not do the mathematical operation of *e* to the *i*th power. However we can substitute θ_3 into Euler's formula, and obain

$$e^{i\theta_3} = \cos\theta_3 + i \sin\theta_3$$

$$\textit{and } z_3 = r_1 r_2 e^{i(\theta_1 + \theta_2)} = r_1 r_2[\cos(\theta_1 + \theta_2) + i \sin(\theta_1 + \theta_2)]$$

Note that the right hand side of the last equation is just the **same** as the polar equation we solved above for point *3*. All three coordinate systems therefore give the same answer.

As another point in the multiplication of complex numbers, note that multiplying a complex number (or real number for that matter) by *i* as the effect of *rotating* its value by *90°*. Let us multiply the value of point 1 by *i* four times, which gives

$$i(2 + 1i) = -1 + 2i \quad (Pt.\ 2)$$

$$i(-1 + 2i) = -2 - 1i \quad (Pt.\ 3)$$

$$i(-2 - 1i) = 1 - 2i \quad (Pt.\ 4)$$

$$i(1 - 2i) = 2 + 1i \quad (Pt.\ 1)$$

And when these values are plotted in Fig. 7-28, we see that each multiplication by *i* just has the effect of rotating point *1* by *90°*. And since *i* to the *4th* power is just equal to *1*, any value (or point address) that is multiplied by *1* returns to the exact same value.

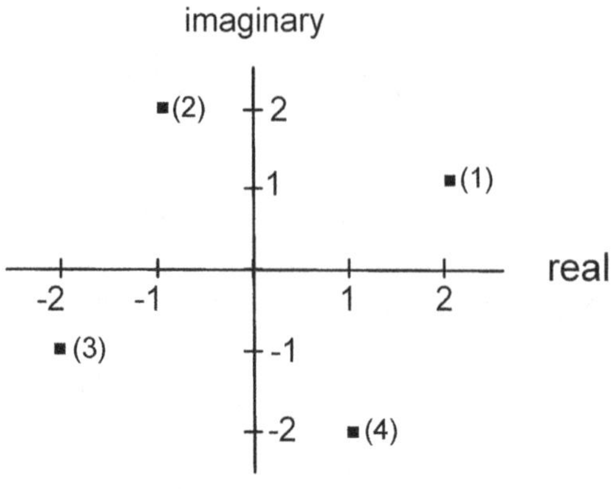

Fig. 7-28 Rotation by multiplying by i

Role of Imaginary *i* in AC Electricity

Our homes use **alternating current (ac)** where the driving **voltage** takes the form of a *sine wave*. If the supply network and household circuit contained only **resistors** (a hypothetical situation), the current would be given by Ohm's law, $V = iR$, and be in perfect phase with the voltage. If we let $R = 2$ *ohms*, and the voltage be *120 volts* (rms), then we would have the *i-V* relationship shown in Fig. 7-29.

Both the voltage and current change constantly as the rotor of the power plant generator turns. And while the household voltage is normally given as *110-120 volts*, this is the **root-mean-square (rms)** value. Note that the peak voltage in Fig. 7-29 is ~170 volts. To find the average voltage over the entire cycle, we multiply the peak voltage × *0.707*, which gives a rms value of *~120* volts. This is the voltage used in electrical calculations.

The **power** used by an electrical device or in a circuit is then defined as

$$P = V_{rms} I_{rms} \cos\varphi$$

Note the factor, "*cosφ*". This is a function of the **difference in phase** between the applied voltage, and the resultant current behavior. For a circuit containing only resistance, there is no difference in phase, and so $\varphi = 0$ and $\cos\varphi = 1$. Therefore, $P = V_{rms} I_{rms}$ for circuits having only resistance.

However all practical electrical circuits are made up of *three* components, these being **resistors**, which are any part of the circuit that resists the flow of current (and generates heat), **capacitors**, which can be viewed as non-connected plates that "store up" electrons, and **inductors**,

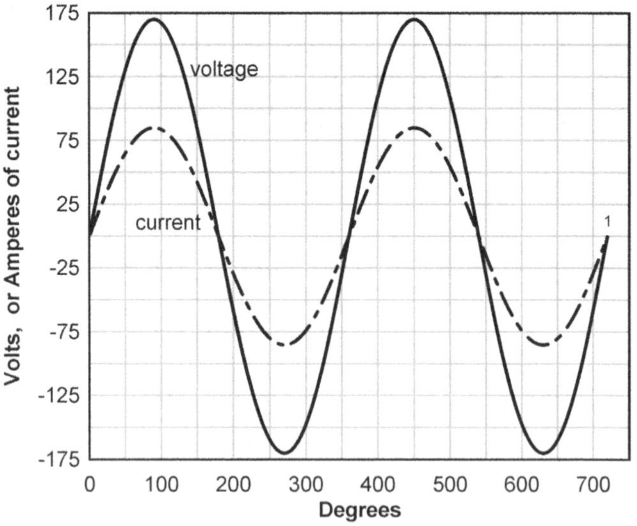

Fig. 7-29 Voltage and current curves in AC electricity

which are effectively coils of wire in which current is induced. In electrical parlance, resistors produce a **resistance** (they resist the flow of the current), whereas capacitors and inductors produce a **reactance** (they "react" with the current).

Electrical measurements using an oscilloscope show that the waveforms of the AC voltage and current are affected differently by the three components:

The voltage across a resistor is **in phase** with the current.

The voltage across a capacitor **lags** the current by 90°.

And, the voltage across an inductor **leads** the current by 90°.

Electrical engineers analyze electrical circuits using *complex numbers* and the imaginary-real number plane. The total voltage across a circuit consisting of resistors, capacitors and inductors is written as $\mathbf{V_{RLC}}$. The **impedance, Z,** of the circuit is the total resistance to current flow offered by the *combination* of all of the circuit components, i.e.,

$$V_{RLC} = IZ, \text{ where } I \text{ is the current through the circuit}$$

If the *reactances* across the inductors and capacitors are defined as *imaginary* (which we indicate here as *j* to avoid confusion with *i*, the symbol for current) it follows that

$$Z = R + j(X_L - X_C)$$

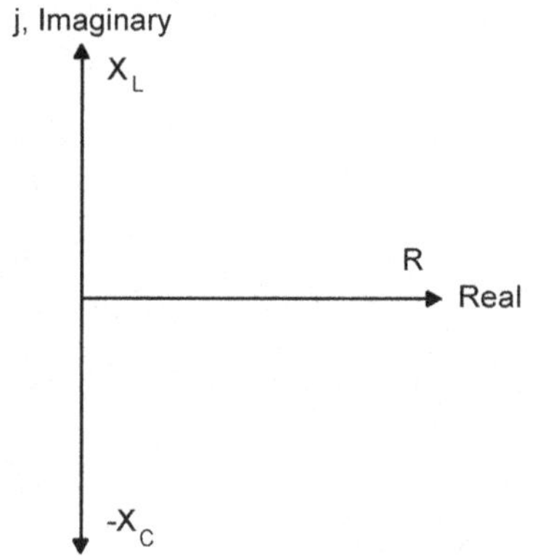

Fig. 7-30 Resistor, Inductor and Capacitor Impedances

Since the current leads with inductance, X_L is plotted as a *positive* imaginary number. And since the current lags with capacitance, X_C is plotted as a *negative* imaginary number. The vector plot of impedances is shown in Fig. 7-30.

If the complex numbers are treated in polar coordinates, we find that the absolute value of **Z**, and the phase angle, φ, of Z with respect to the real axis are given by, respectively,

$$|Z| = \sqrt{R^2 + (X_L - X_C)^2} \quad \textit{and} \ \tan\varphi = \frac{X_L - X_C}{R}$$

Let us calculate the impedance of a circuit which has a resistor of 5 ohms *in series* (parallel circuits are more complex) with an inductor of a reactance of 4 ohms.

$$|Z| = \sqrt{R^2 + (X_L - X_C)^2} = \sqrt{5^2 + 4^2} = \sqrt{41}; \ Z = 6.40$$

$$\textit{and} \ \tan\varphi = \frac{X_L - X_C}{R} = \frac{4}{5} = 0.80; \ \varphi = 38.66^o$$

The angle, φ [illegible] angle [illegible] the [illegible] voltage, and φ is positive, so the voltage *leads* the [illegible] *38.66°*.

Assume the [illegible] ns the [illegible] comp[illegible] ries: resistor (5 ohms), inductor (3 ohms) and capaci[illegible]). Th[illegible] lati[illegible] pedance of the circuit is

$$|Z| = [illegible] - X [illegible] + (3 [illegible] \sqrt{50}; \quad Z = 7.07$$

$$[illegible] = \frac{X_L - [illegible]}{R} [illegible] \frac{[illegible]}{5} \quad \varphi = -45.00^o$$

In this case, φ is eq[illegible] which [illegible] that [illegible] *lags* the current by *45.00°*. If plotted, the vector [illegible] nd d[illegible] the [illegible] ant of the reference circle at an angle of *minus 45.0* [illegible] real [illegible]

Significan[illegible] phase [illegible]

In electrici[illegible] and [illegible] the [illegible] between the current and voltage is of vital i[illegible] Engin[illegible] thre[illegible] ts in AC flow: **(P)**, the "real" power that perform[illegible] k, **(S)** [illegible] nt p[illegible] the volt-amperes (VA) of power supplied to t[illegible] **Q)**, t[illegible] p[illegible] consumed in *non-productive* capacitance and in[illegible] ions i[illegible] it.

These comp[illegible] defin[illegible] fa[illegible] **tor triangle** of the circuit in question:

$$[illegible]\textit{er Fa}[illegible] \frac{\textit{real}}{\textit{appar}[illegible]} = \frac{P}{S}$$

$$[illegible]\textit{Vecto}[illegible] \quad Q^2$$

That is, S is the pow[illegible] the [illegible] par[illegible] ver used for work, and Q is the part of power wast[illegible]

In accord wi[illegible] dex m[illegible] lat[illegible] we have that, if φ is the phase angle between the [illegible] oltag[illegible] have [illegible] taken φ as being the angle between the *voltag*[illegible], the p[illegible] or [illegible] os φ|, and

[illegible] cos [illegible]

We can see how th[illegible] curr[illegible] reduce the "instantaneous" power by comparin[illegible] with Fi[illegible] Fi[illegible] re the voltage and current are in perfect phase, the instantaneous power at *90º* is given by P = V (*170* volts) × *i* (*170* volts/*2* ohms = 85 amperes), or *170* volts *x 85* amperes = ***14.450* kW**.

In Fig. 7-31 where the voltage leads the current by *45º*, the peak voltage (*170* volts) occurs at *45º* of the cycle, but the current at this point is only about *60* amperes, and the instantaneous power is *170* volts *x 60* amperes, or ***10.200* kW**. At the amperage peak at 90º, the current is *85* amperes and the voltage is 120 volts. The instant power at this point is *85* amp × *120* volts = *10.200* kW.

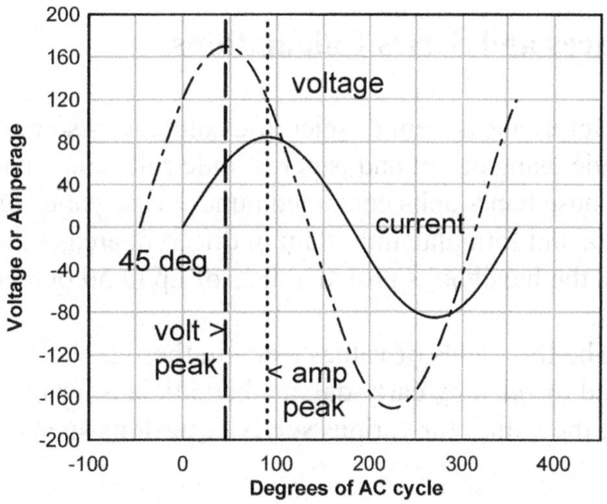

Fig. 7-31 Loss of power by 45 deg leading voltage

Comparing all of Fig. 7-31 with Fig. 7-29 shows that there is *no* point in the AC cycle of Fig. 7-31 where the instantaneous power is as high as in Fig. 7-29. Having the voltage lead (or lag) the current therefore reduces the real power at all points in the AC cycle.

Moreover, the amount of power that is available for work when φ is equal to *45º* is predicted by

$$P = S\,|\cos\varphi|,\ and\ P = 14.450\ kW \times \cos 45^{o}(= 0.707) = 10.216\ (\sim 10.200)\ kW$$

For a circuit having a phase angle of *45º*, one must therefore supply *1.42 kW* to the load site to obtain *1 kW* of actual "working" power. An extra *29.6%* of "waste" power must be generated and transported to the load site, which requires the expense of larger generators, wires, and transformers, as well as greater fuel costs.

To improve the load factor, many plants and large-scale electricity consumers install **power factor correction units**. These are electronic controller devices which usually have a number of capacitors that can be switched as needed into the master circuit. This reduces the sum of the *reactance vectors*, X_L - X_C (see Fig. 7-30), and moves the phase angle toward *zero* (*cos 0º = 1*) which improves the power factor.

Chapter 8.

Sequences and Series Calculations

Before the advent of scientific calculators some *40* years ago, mathematical calculations were made using pencil and paper, a slide rule, and a handbook containing "trig" and "log" tables. These handbooks contained hundreds of pages of the various math functions. Numerous other data and formulae important in different areas of mathematics were also given. For calculus, the handbooks provided lists of up to *50* or so differentials, and over *700* specific integrals.

The thousands of values listed in these tables (most to *5* decimal places or better) were calculated *by hand* by early-day mathematicians, or by "computers" who were persons hired to compute the values for various **series equations** such as

$$\sin x = x - \frac{x^3}{3!} + \frac{x^5}{5!} - \frac{x^7}{7!} + \cdots$$

These equations are actually approximations, which can disturb those who want the "real" answer. But recall that *all* irrational numbers including π and e are just approximations. We may know their value to millions of places, but we never know them exactly. Their values can be calculated accurately enough, however, for all practical purposes.

When looking at a series equation, one's first thoughts are "How do they know what terms to put into the series?", and "How do they know that the sum of the series will come out to be the right answer?" Mathematicians have studied these questions for hundreds of years. In this Chapter, we will show how these two questions have been answered for functions such as $\sin\theta$, $\cos\theta$, e^x, $\ln x$, a^x and others.

The literature concerning sequences and series is large, and we will look only at a very small part here. We will concentrate on the **power series**, **Newton's Method for Solving Equations**, and the **Taylor series**. These have the great advantage that they can be used to solve complex functions by approximation in a straightforward way. It is not easy, for example, to find the value of the function $\sin x$ directly. But if we establish that $\sin x$ is equal to the *series* function,

$$x - \frac{x^3}{3!} + \frac{x^5}{5!} - \frac{x^7}{7!} + \cdots$$

then $\sin x$ can be calculated in a routine manner that makes the solving of $\sin x$ for incremental values of x such as *0.510, 0.511, 0.512 ...* radians relatively easy.

We will also examine the **Fibonacci sequence**, which is a well-known sequence from mathematical folklore, and the **Machin-like formulae**. The Machin-like formulae involve a series, but one with only a *finite* number of terms as opposed to regular series where the number of terms is infinite. Computer programs which can calculate π to many millions of decimal places are based on Machin-like formulae.

Sequences

Sequences and series are interrelated. A **sequence** is just a list of numbers that occur in some *specific* order. Sequences are enclosed in *braces*, and are given in such forms as

$$\{ a_1, a_2, \cdots a_n, a_{(n+1)} \cdots \} \qquad \{ a_n \} \qquad \{ a_n \}_{n=1}^{\infty} \qquad \{ a_n \}_{\geq 1}$$

The subscript, $n = 1$, indicates that the terms begin with n equal to the number 1, and goes up by *integers* (no fractions). The superscript, ∞, indicates that n can become infinitely large. In the fourth term, the subscript ≥ 1 says that n must be an integer equal to, or greater than 1.

The specific order of the numbers in the sequence is usually given by a formula, which is either known beforehand, or is determined by the analysis of an "experimental" sequence. One example of a formula for producing a sequence is

$$\left\{ \frac{n+1}{n^2} \right\}_{n=1}^{\infty} = \frac{2}{1}, \frac{3}{4}, \frac{4}{9}, \frac{5}{16}, \frac{6}{25}, \frac{7}{36}, \cdots \frac{11}{100}, \frac{12}{121}, \cdots \frac{101}{10{,}000}, \cdots$$

The principal mathematical property of a sequence is its **limit**, which is the number that the *final* term of the sequence approaches as $n \to \infty$.

The terms in this particular sequence approach zero, but slowly. For $n = 1000$, the value of the term is $1001/1{,}000{,}000$ or ~ 0.001. If the exponent in the denominator is larger, say 4, then the value at $n = 10$ becomes $11/10{,}000$ or ~0.001. The second expression approaches 0 at a rate 100 times greater than the first expression. In either case, the sequence *has* a limit, and that limit is zero regardless of how fast it is approached. This is expressed in equation form as

$$\lim_{n \to \infty} \left\{ \frac{n+1}{n^2} \right\}_{n=1}^{\infty} = 0$$

The ordering of the sequence can be given by other formulas, and the nature of the limit will depend upon the given formula, for example,

$$sequence = \left\{ \frac{3n^3 + 6}{4n^2} \right\} \; and \; \lim_{n \to \infty} \left\{ \frac{3n^3 + 6}{4n^2} \right\} = \infty$$

The limit of this se[illegible] [illegible]inity be[illegible] [illegible] n^3 [illegible] [illegible]umerator overrides the n^2 term in the denominator [illegible] [illegible] the co[illegible] [illegible] [illegible] goes to infinity. The value of the final term in the [illegible] [illegible]erefor[illegible] [illegible]nfi[illegible] [illegible]act to a *positive* infinity.

If it should [illegible] [illegible] the fo[illegible] [illegible] [illegible] its limit would be *negative* infinity,

$$[illegible] = -\{[illegible] \; and! \; [illegible] \frac{[illegible]\; 6}{[illegible]}\} = -\infty$$

We can als[illegible] [illegible] [illegible]ence [illegible] [illegible]ndiv[illegible] [illegible] alternate between being positive and negati[illegible] [illegible] [illegible] the pr[illegible] [illegible] the [illegible] [illegible]rmula,

sequ[illegible] [illegible] $1)^n \}$, [illegible] $\{-1\}^n$ [illegible] *undefined) limit*

In this case, there i[illegible] [illegible] [illegible]cause [illegible] [illegible]ust [illegible] value, and not a set of alternating values.

Finally, the [illegible] [illegible] [illegible]quence [illegible] [illegible] [illegible]ber. Consider this formula for a sequence,

$$sequ[illegible] \; \frac{4[illegible] + [illegible]}{[illegible] - 3n^2} \; [illegible] \; \frac{[illegible]n^2}{[illegible]n + [illegible]} \; \frac{\left(4 + \frac{4}{n^2}\right)}{\left(3 + \frac{6}{n}\right)} = \frac{4}{3}$$

By factoring n^2 fro[illegible] [illegible] [illegible]umera[illegible] [illegible] [illegible]om[illegible] [illegible]oting that any integer divided by n (or its powers) [illegible] [illegible] 0 whe[illegible] [illegible] in[illegible] [illegible]d that the sequence limit is *4/3*.

Any sequen[illegible] [illegible] [illegible]s a lim[illegible] [illegible] ∞ ([illegible] [illegible]ot defined numbers), or which consists of number[illegible] [illegible] [illegible]ng sig[illegible] [illegible]ot a [illegible] [illegible]er) is said to **diverge**. Any sequence which ha[illegible] [illegible] [illegible]r of [illegible] [illegible]num[illegible] [illegible]o **converge**. The first and last sequences above ar[illegible] [illegible] [illegible]onver[illegible] [illegible] th[illegible] [illegible]e sequences are divergent.

Fibonacci [illegible]

The Fibona[illegible] [illegible] [illegible], or "Fi[illegible] [illegible] [illegible]umb[illegible] [illegible]aptured much public interest in recent years. A re[illegible] [illegible] [illegible]any [illegible] [illegible] [illegible] [illegible] [illegible]oncerning Fibonacci numbers is given by R. Knott [illegible] [illegible] [illegible]rre[illegible] [illegible] [illegible]ons[illegible] [illegible]ibonacci/

The Fibona[illegible] [illegible] [illegible]s di[illegible] [illegible]use [illegible] [illegible]**sive** sequence. That is, the value of a given nu[illegible] [illegible] [illegible]sequen[illegible] [illegible]mi[illegible] [illegible]*us* numbers in the sequence. (Cursive relates to [illegible] [illegible] [illegible]cursi[illegible] [illegible] [illegible]ough [illegible]ning "curve (or turn) back on.")

The Fibonacci sequence is named after an Italian mathematician who proposed a problem ca. *1202* involving the growth of a population of rabbits. The question was, starting with one pair, how many pairs of rabbits would be present at the end of *each* month, if the gestation period is one month, and each pair must mature one month before breeding.

The problem can be approached in this way: Starting with conception at *time* = *0*, one pair (**Pair** ***1***) of rabbits is born at the end of month *1*. But this pair must mature for *1* month before it can reproduce. At the end of month 2, the **Pair** ***1*** matures and then breeds . At month *3*, **Pair** ***2*** is born, so there are now *2* pairs of rabbits. At month *4*, the **Pair** ***1*** produces **Pair** ***3***, but **Pair** ***2*** is not mature, so there are only *3* pairs of rabbits. At month *5*, **Pair** ***1*** produces **Pair** ***4***, and **Pair** ***2*** produces **Pair** ***5*** (**Pair** ***3*** is immature), so we have *5* pairs of rabbits total. At month *6*, **Pair 1** produces **Pair** ***6***, **Pair** ***2*** produces **Pair** ***7***, and **Pair** ***3*** produces **Pair** ***8***. But **Pairs** ***4*** and ***5*** are not yet mature, so only *8* pairs of rabbits result. At month *7*, **Pair** ***1*** produces **Pair** ***9***, **Pair** ***2*** produces **Pair** ***10***, **Pair** ***3*** produces **Pair** ***11***, and **Pairs** ***4*** and **5** (now mature) produce **Pairs** ***12*** and ***13***. Therefore, we have *13* pairs of rabbits at month *7*, with **Pairs** ***6, 7,*** and ***8*** being immature. This process is repeated for months n = 8, 9, 10 ...

Such a "brute force" calculation soon becomes tedious. But we see that the sequence

$$1,\ 1,\ 2,\ 3,\ 5,\ 8,\ 13\ \cdots$$

has the property that F(1) + F(2) = F(3). That is, $1 + 1 = 2,\ 1 + 2 = 3,\ 2 + 3 = 5,\ 3 + 5 = 8,$ and so on. This recursive procedure can be used to generate the "Fibonacci" sequence for as many terms as we may desire, i.e.,

$$1,\ 1,\ 2,\ 3,\ 5,\ 8,\ 13,\ 21,\ 34,\ 55,\ 89,\ 144,\ 233,\ \cdots$$

The Fibonacci sequence holds a fascination for mathematicians principally because (1) it appears to mirror growth processes in some plants and seashells and (2) the ratio, F(n)/F(n-1), approaches the number, *1.61803...*, as *n* becomes large. The number, *1.61803...*, is a special number known to the ancient Greeks as **phi (φ), the Golden Ratio**. It was thought to have mystical power, and the Greeks believed that the sides of any rectangular form in art or architecture should be in the *golden ratio* for maximum beauty and style.

In mathematical terms, the Golden Ratio has the special property that it can be defined in terms of itself, i.e.,

$$\varphi = 1 + \frac{1}{\varphi}$$

If we multiply both sides of this equation by φ and rearrange, we obtain a quadratic equation which can be solved by the quadratic formula,

$$\varphi^2 - \varphi - 1 = 0, \quad \textit{and by the quadratic formula,} \quad \varphi = \frac{\pm (1 + \sqrt{5})}{2} = 1.61803...$$

$$\textit{and with the other root} = -0.61803...$$

where one root of the quadratic equation is *1.61803... (=* φ*)*, and the *other* root is -0.61803...

One use of the golden ratio, φ, occurs in **Binet's formula** (**J. Binet**, *1786-1856*) below, which can be used to calculate the Fibonacci number for *any* n,

$$F(n) = \frac{\varphi^n - (1 - \varphi)^n}{\sqrt{5}}$$

Let us calculate the 12th term in the Fibonacci sequence which is *144*,

$$F(n) = \frac{\varphi^{12} - (1 - \varphi)^{12}}{\sqrt{5}} = \frac{(1.61803...)^{12} - (-0.61803...)^{12}}{2.23606...}$$

$$= \frac{(321.99689... - 003.10537...)}{2.23606...} = 144.00051...$$

The answer comes out to be essentially equal to the integer, *144*, which is the 12th term of the Fibonacci sequence.

Let us look at one derivation of Binet's formula as an example of mathematical logic. Our approach is inspired by that of M. Anderson, J. Frazier and K. Popendorf at the web site *http://library.thinkquest.org/27890/applications2p.html*. Notice that φ is one root of the quadratic equation, $\varphi^2 - \varphi - 1$. It's value is *1.61803...* But the equation also has a second root which is $1 - \varphi$, or *1 - 1.61803 = -0.61803*. We will call this second root *psi or* ψ. We can therefore rewrite Binet's formula as

$$F(a) = \frac{1}{\sqrt{5}} (\varphi^a - \Psi^a)$$

Going back to the original quadratic equation, we see that it can be put into the form of an *identity*, that is

$$\varphi^2 = \varphi + 1$$

If we multiply both sides of this equation repeatedly by φ, and reapply the identify, $\varphi^2 = \varphi + 1$, in the resultant equations, we obtain this sequence

(*original identity*)	$\varphi^2 = 1\varphi + 1$
$\varphi^1\varphi^2 = \varphi(\varphi + 1) = \varphi^2 + \varphi = (\varphi + 1) + \varphi;$	$\varphi^3 = 2\varphi + 1$
$\varphi^1\varphi^3 = \varphi(2\varphi + 1) = 2\varphi^2 + \varphi = 2(\varphi + 1) + \varphi;$	$\varphi^4 = 3\varphi + 2$
$\varphi^1\varphi^4 = \varphi(3\varphi + 2) = 3\varphi^2 + 2\varphi = 3(\varphi + 1) + 2\varphi;$	$\varphi^5 = 5\varphi + 3$
$\varphi^1\varphi^5 = \varphi(5\varphi + 3) = 5\varphi^2 + 3\varphi = 5(\varphi + 1) + 3\varphi;$	$\varphi^6 = 8\varphi + 5$
$\varphi^1\varphi^6 = \varphi(8\varphi + 5) = 8\varphi^2 + 5\varphi = 8(\varphi + 1) + 5\varphi;$	$\varphi^7 = 13\varphi + 8$
$\varphi^1\varphi^7 = \varphi(13\varphi + 8) + 13\varphi^2 + 8\varphi = 13(\varphi + 1) + 8\varphi;$	$\varphi^8 = 21\varphi + 13$

The coefficients in this sequence are just the Fibonacci numbers after the starting numbers of *1,1*. For example, the *8th* number in the Fibonacci sequence is *21*, and the *7th* number is *13*. We have therefore that

$$\varphi^8 = F(8)\varphi + F(8-1), \textit{ or generalizing, } \varphi^a = F(a)\varphi + F(a-1)$$

However, if we evaluate this equation using $\varphi = 1.61803...$, we find that it gives

$$(\varphi^8 =)\ 46.97778... \approx (21\varphi + 13 =)\ 46.97863...$$

The numerical value of this equation is approximately equal and mathematically correct. But *46.97...* is *not* the value of the *8th* Fibonacci number, or is it even a Fibonacci number. Our difficulty lies in the fact that this is only a *step* in the derivation of Binet's formula. We must proceed further.

The equation

$$\varphi^8 = F(8)\varphi + F(8-1), \textit{ or } \varphi^a = F(a)\varphi + F(a-1)$$

is just a version of the original quadratic equation, $\varphi^2 = \varphi + 1$, and the relationship holds because φ is a root of that equation. But we recall that ψ is also a root of the quadratic equation, and therefore we can write

$$\Psi^a = F(a)\Psi + F(a-1)$$

If we subtract the two equations, the $F(a - 1)$ terms subtract out, and the final equation is

φ^a [illegible] $(a)(\varphi$ [illegible] or $F(a) = \dfrac{\varphi^a - \Psi^a}{\varphi - \Psi}$

Since φ = [illegible] ψ = [illegible] the [illegible] $- \psi = 2.23606...$ which is the square root of 5. A[illegible] o th[illegible] [illegible]ere $a = n$, we have Binet's formula.

[illegible]

Fibonacci [illegible] and [illegible]

In the rabbi[illegible] problem [illegible] simply reflect a growth process which occurs expo[illegible] [illegible] d be

$F(n$ [illegible] e [illegible]

where the growth c[illegible] ln of [illegible] ratio φ) between $F(n)$ and $F(n-1)$,

$\dfrac{F(n)}{F(n-1)}$ = [illegible] growth constant [illegible]033989) = 0.481211825

We should therefore [illegible] calculat[illegible] [illegible] of numbers by the exponential growth equation. [illegible] ratio of [illegible] [illegible]s φ only at the larger Fibonacci numbers, we may [illegible] conv[illegible] [illegible] at the **decay** equation.

The decay [illegible] same [illegible] growth [illegible] except the decay constant is the negative (-0.48121[illegible] growth [illegible] calculator, we can then generate the Fibonacci numbers [illegible]

$F(n$ [illegible] $e^{-0.481}$ [illegible] starting [illegible] = 144, *we have*

Fibonacci	144	[illegible]	[illegible]	[illegible]	[illegible]	[illegible]	[illegible]	5	3	2	1	1
Calculated	[illegible]	[illegible]	[illegible]	[illegible]	[illegible]	[illegible]	[illegible]	4.96	3.06	1.89	1.17	0.72

And we see good agreement except at the very lowest numbers.

If we wished to find not the next Fibonacci number down, but the *4th* number down, we would modify the equation to the form,

$$F(n-4) = 144\ e^{(4 \times -0.481211825)} = 21.01 \ (4th\ term\ from\ 144)$$

To find the *4th* term up, which is *F(16)*, we would set the exponent to *(4 x 0.481211825)* and find *F(16)* to be *986.99... (i.e., 987).*

Because of the inaccuracy in the decimal calculations, the numbers do not come out as exact integers. But since we know we are dealing with "whole" rabbits, we know to round to the integer. It is, in fact, a *general property* of sequences that they deal only with integers, or in some cases, fractions which are one integer divided by another.

The Internet contains many articles concerning Fibonacci numbers, and their occurrences in unexpected places. However, sequences in general play a relatively small part in applied mathematics, and we will not study them farther in this tutorial.

Series

Students sometimes confuse sequences and series. Consider the sequence

$$\left\{ \frac{1}{n} \right\}_{n=1}^{\infty} = \frac{1}{1}, \frac{1}{2}, \frac{1}{3}, \frac{1}{4}, \ldots, \frac{1}{n}$$

The **limit of the sequence** as *n* goes to *infinity* is *0*. As the value of *n* becomes infinitely large, the value of the final term, *1/n*, must approach zero But let us transform this sequence into a **series**,

$$\sum_{n=1}^{\infty} \left(\frac{1}{1} + \frac{1}{2} + \frac{1}{3} + \frac{1}{4} + \cdots \right) \quad or\ in\ general, \ \sum_{n=1}^{\infty} \frac{1}{n}$$

In a series, the individual numbers (or terms) are not just listed, but rather are *summed* (this is indicated by the preceding Σ symbol) for all values of *n* up to *infinity*. Note also that a series is usually inclosed by parentheses.

This brings up the question of what is the **limit of the series**, or in other words, what would all of the terms together add up to. The final term of the series becomes "infinitesimally" small as *n* becomes larger and larger, but all of the preceding terms have some numerical value. One might guess that the sum of terms would approach some *finite* limit. But this is not true.

The actual limit of the series is *infinity*, which is written in equation form as

$$\lim_{n\to\infty} \sum_{n=1}^{\infty} \frac{1}{n} = \infty \quad \textit{or also as} \quad \lim_{n\to\infty} \left(1 + \frac{1}{2} + \frac{1}{3} + \cdots \frac{1}{n}\right) = \infty$$

That the limit of the series is infinity can be shown with some use of calculus (see Chap. 9 on calculus). First let us consider the *variable*, *x*, where a function of *x*, which we write as *f(x)*, is equal to *1/x*, and *1/x* is a derivative, i.e.,

$$f(x) = \frac{1}{x}, \quad \textit{for } x \geq 1, \quad \textit{and } \frac{1}{x} \textit{ is a } \underline{\textit{derivative}}$$

Since the function, *f(x)*, is a derivative, we can integrate the function and obtain the integral, which is the *natural log function*. This integral can be evaluated from *n* to *n+1* as

$$\int_{n}^{n+1} \frac{1}{x} dx = [\ln(x)]_{n}^{n+1} = \ln(n + 1) - \ln(n)$$

If we integrate starting from *1*, then *n= 1*, and *ln(1) = 0*, so the value of the integral is just equal to *ln(n+1)*,

$$\int_{1}^{n+1} \frac{1}{x} dx = [\ln(x)]_{1}^{n+1} = \ln(n + 1) - 0, \quad \textit{and so } \lim_{n\to\infty} \ln(n+1) = \infty$$

This is because, as *n+1* becomes infinitely large, so must the expression, *ln(n+1)*, and therefore the limit of *ln(n+1)* is equal to ∞.

Now we must connect the integer series with the natural logarithm integral. To do this, we note that for *any x* there are two *integers*, *n* and *n+1*, such that *x* lies between them. And so we can write

$$\frac{1}{n + 1} \leq \frac{1}{x} \leq \frac{1}{n}$$

If we integrate each of these *3* terms from *1* to *n + 1*, we have

$$\int_1^{n+1} \frac{1}{n+1} dx \leq \int_1^{n+1} \frac{1}{x} dx \leq \int_1^{n+1} \frac{1}{n} dx$$

For this proof, we need to be concerned only with the *1/n* integer series. Note that *n* is a smaller number than *x*; hence *1/n* is larger than *1/x*. Since *1/n* is a *constant*, its integral follows the integration rule for constants, i.e.,

$$\int a dx = ax, \quad so \left(\frac{1}{n}\right)\int dx = \left(\frac{1}{n}\right) x = \left(\frac{1}{n}\right) x \,\Big|_1^{n+1}$$

We can reproduce the integral from *1 to n+1* by summing the integrals of the "micro-intervals" from *n =1 to n =2, n=2 to n=3, n=3 to n=4*, and so on. This gives the following series, where the value of the *constant* is in the denominator, and the value of the *integral* in the numerator,

$$\left(\frac{1}{n}\right) x \,\Big|_1^{n+1} = \textit{the series, } \frac{1}{1}\left(=\frac{2-1}{1}\right) + \frac{1}{2}\left(=\frac{3-2}{2}\right) + \frac{1}{3}\left(=\frac{4-3}{3}\right) + \cdots + \frac{1}{n}$$

$$= 1 + \frac{1}{2} + \frac{1}{3} + \frac{1}{4} + \cdots + \frac{1}{n}$$

This is just the same series as our original *1/n* series for which we seek the limit as $n \to \infty$. And this series is equal to, or l*arger* than, the integral for *dx/x* as we see here,

$$\frac{1}{n+1} \textit{ series} \leq \int_1^{n+1} \frac{1}{x} dx \; (= \ln(n+1)) \leq 1 + \frac{1}{2} + \frac{1}{3} + \cdots + \frac{1}{n}$$

So it must follow that

$$\textit{if } \lim_{n\to\infty} \ln(n+1) = \infty, \textit{ then so must } \lim_{n\to\infty}\left(1 + \frac{1}{2} + \frac{1}{3} + \cdots + \frac{1}{n}\right) = \infty$$

Mathematical reasoning thus shows the *counter-intuitive* fact that the limit of the *1/n* series (which is known as the **harmonic series**) must go to infinity as *n* goes to infinity. The approach to infinity is very slow, however, with more than 10^{43} terms being required to reach a sum of *100* for the [illegible], *Ca*[illegible] *E*[illegible]), S. P. Thompson and M. Gardner. St. Martin[illegible] [illegible] York [illegible]

Although w[illegible] pursue [illegible], [illegible]ation of the limit of a series is an important part o[illegible], and [illegible] math textbooks and on the Internet (see, e.g., [illegible] *Ma*[illegible] [illegible] *& Sequences*).

Series Cal[illegible]

Series calcul[illegible] importan[illegible] the [illegible] *most precise* calculation of trigonometric, loga[illegible] certai[illegible] able [illegible]. Reconsider the equation

$$[illegible]\ x = [illegible] \frac{x^5}{5!} \ldots$$

Let us calculate the [illegible] *radian*[illegible]). [illegible]or gives the answer to be *0.932039086*. And [illegible] valu[illegible] that

$$0.932039086 \ [illegible]\ \frac{(1.2)^3}{3!} \quad \frac{(1.2)^5}{5!} \quad \frac{(1.2)^7}{7!} + \frac{(1.2)^9}{9!} - \ldots$$

$$- 0.2[illegible] \quad [illegible]207[illegible] \quad [illegible]007109_5 + 0.00001421_9$$

$$[illegible]912 \quad [illegible]736 \quad [illegible]02505 \quad 0.932039269$$

Summing up the te[illegible] [illegible]ght sid[illegible] equati[illegible] that by the *5th* term we have obtained the value [illegible] *radians* [illegible] *5th* d[illegible]e. This series therefore converges quickly [illegible] correct [illegible] [illegible]o more terms would bring us to complete agreemen[illegible] [illegible]al pla[illegible] s th[illegible] that most calculators display. And many calculat[illegible] the [illegible] to c[illegible] *x* values.

Machin-L[illegible] [illegible]tions

Series of th[illegible] are k[illegible] infinit[illegible] cause the number of terms can go to infinity. Cert[illegible] [illegible]ies, c[illegible] series [illegible] a limited number of terms. One of the most im[illegible] se fin[illegible] was [illegible]6 by **J. Machin** (*1680-1751*) for the calculation [illegible] r's fo[illegible] t is,

$$[illegible] = 4\ \text{arc}[illegible] - \text{a}[illegible]\Big)$$

This formula can be developed using only trigonometry. Students wonder why the left side of the equation is given as $\pi/4$. This is because $\pi/4$ in radians is equal to 45°, which has a tangent equal to *1*. It also provides a means for introducing π, which we are seeking to calculate, into the equation.

Machin started with the angle, β, which he proposed to have a tangent of *1/5*, or

$$\tan \beta = \frac{1}{5}$$

He then made use of the standard formula from trigonometry for the tangent of a *doubled* angle, which gave

$$\tan 2\beta = \frac{2 \tan \beta}{1 - \tan^2 \beta} = \frac{2 \times \frac{1}{5}}{1 - \frac{1}{5^2}} = \frac{\frac{2}{5}}{\frac{24}{25}} = \frac{2}{5} \times \frac{25}{24} = \frac{5}{12}$$

Machin subsequently applied the tangent double-angle formula again, and found

$$\tan 4\beta = \frac{2 \tan 2\beta}{1 - \tan^2 2\beta} = \frac{2 \times \frac{5}{12}}{1 - \left(\frac{5}{12}\right)^2} = \frac{\frac{10}{12}}{\frac{119}{144}} = \frac{10}{12} \times \frac{144}{119} = \frac{120}{119}$$

Machin recognized that the tangent for 4β of *120/119* was only *1/119* above the tangent of $\pi/4$, which is *1*. He cast this difference in the form of the standard trigonometric equation for the subtraction of the tangents of two angles, i.e.,

$$\tan (a - b) = \frac{\tan a - \tan b}{\tan b + \tan a}$$

Since angle a = angle 4β, and angle b = angle $\pi/4$, and also $\tan \pi/4 = 1$, we then have that

$$\tan\left(4\beta - \frac{\pi}{4}\right) = \frac{\tan 4\beta - 1}{1 + \tan 4\beta} = \frac{\frac{120}{119} - 1}{1 + \frac{120}{119}} = \frac{1}{239}$$

If we take the *arctan* of both sides of this last equation, this yields

$$4\beta - \frac{\pi}{4} = \arctan \frac{1}{239}$$

And recalling that $\tan \beta = 1/5$, which means that $\beta = \arctan 1/5$, and rearranging the equation, we end up finally with the Machin formula,

$$\frac{\pi}{4} = 4 \arctan\left(\frac{1}{5}\right) - \arctan\left(\frac{1}{239}\right)$$

This simple formula turns out to be remarkably accurate. If we compare π as given on my calculator with π as calculated from Machin's formula, we see excellent agreement,

Calculator π = 3.141592654

Machin formula π = 3.141592654

Interestingly, the literature reports that three other "arctan" two-term formulas for π were found by early-day mathematicians, these being

$$\frac{\pi}{4} = \arctan\left(\frac{1}{2}\right) + \arctan\left(\frac{1}{3}\right); \quad \pi = 3.141592654$$

$$\frac{\pi}{4} = 2\arctan\left(\frac{1}{2}\right) - \arctan\left(\frac{1}{7}\right); \quad \pi = 3.141592654$$

$$\frac{\pi}{4} = 2\arctan\left(\frac{1}{3}\right) + \arctan\left(\frac{1}{7}\right); \quad \pi = 3.141592654$$

Machin did not have a calculator to determine arctan values; rather he used a series expression developed by **J. Gregory** (*1638-1675*), viz.,

$$\arctan(x) = x - \frac{x^3}{3} + \frac{x^5}{5} - \frac{x^7}{7} + \frac{x^9}{9} - \ldots.$$

The value of the terms in this series for the first segment of the Machin formula, where $x = 1/5$, can be calculated and summed as

Term		*Value*
$1/5$	=	0.20000000000000...
$(1/5)^3/3$	=	-0.00266666666666...
$(1/5)^5/5$	=	0.00006400000000...
$(1/5)^7/7$	=	-0.00000182851714...
$(1/5)^9/9$	=	0.00000005688888...
$(1/5)^{11}/11$	=	-0.00000000018681...
$\arctan\left(\frac{1}{5}\right)$	=	0.19739556151827...
		and so on

There is no limit to the number of terms that can be taken, nor to the decimal places to which each term may be calculated. The same treatment can be applied to determine *arctan (1/239)*, the second element of Machin's formula. The ultra-high precision values for the two arctan terms can then be substituted into the original Machin formula, so the degree of precision that can be obtained for π depends only on the energy of the mathematician. Machin himself computed π correctly to *100* places.

Machin-type formulas continue to be developed even up to this day, and were used in the Japanese supercomputer study which calculated π to *1.24 trillion* places. One recent version of the Machin formula from 2003 uses seven, rather than just two, arctangent terms, and the denominator of the last term is not *239*, but rather a very large number of *20 digits*.

Power Series

One of the classic examples of a "series" problem involves a situation where some quantity is divided by ½ repeatedly. Say you make *$1* the first day of labor, *$0.50* the next day, *$0.25* the third day, and so on. How much money would you make by working an infinite number of days? The answer can be given in series form, where *s* is the **sum total** of the series,

$$s = 1 + \frac{1}{2} + \frac{1}{4} + \frac{1}{8} + \frac{1}{16} + \cdots$$

If we multiply each side of the equation by *1/2*, we obtain a second equation

$$\frac{1}{2}s = \frac{1}{2} + \frac{1}{4} + \frac{1}{8} + \frac{1}{16} + \cdots$$

And if we subtract [illegible] ation [illegible] the sum of the series is equal to *2*, or in other words, [illegible] *converg*[illegible]

$$\frac{1}{2}s = 1, \quad [illegible] \quad 2$$

You would earn $2 [illegible] infini[illegible] ed for an infinite number of days under the wag[illegible] pose[illegible] ial p[illegible]

This solutio[illegible] ten fo[illegible] al te[illegible] ded that q is *less* than *1* (if q is greater than *1* the se[illegible] erge), i[illegible]

$$[illegible] \; q \; [illegible] \; + \cdots$$

$$or \; [illegible] \; s \; above \; [illegible] \left(\frac{1}{2}\right) [illegible] + \left(\frac{1}{2}\right)^3 + \cdots$$

If we do the same [illegible] and [illegible] pro[illegible] ove, we end up with an expression for the s[illegible] ies, [illegible]

$$s \; [illegible] \; and \; [illegible] \; \frac{1}{1 - [illegible]} \; [illegible] \textit{eneral solution})$$

For various values [illegible] $q =$ [illegible] $q =$ [illegible] on, the sum of the series can be determined simply [illegible]

$$s = \frac{[illegible]}{1 [illegible]} = \frac{1}{1 - [illegible]} = \frac{1}{1 - [illegible]} = \frac{\textit{denominator of } q}{(\textit{denom. of } q) - 1}$$

Note that q [illegible] to be [illegible] atio [illegible] 33, $s = 1/0.667$ or 1.499 (≈ 1.50 as above). If $q = 0.$[illegible] $1/0.7$[illegible] abo[illegible] s will always be just the reciprocal of the de[illegible] $1 - q$) [illegible] hat [illegible] as q is less than *1*.

In some cas[illegible] may [illegible] ed [illegible] the terms of the series are multiplied by a *con*[illegible] One [illegible] determ[illegible] mmon factor, and then the total sum (or limit) of th[illegible] omes j[illegible]

$$[illegible] = s \; [illegible] \; fi[illegible] \; \frac{1}{[illegible] - q}\Big)$$

If we substitute x for q, and introduce the series value for s, we obtain the **geometric power series**, i.e.,

$$\frac{1}{1 - x} = s = 1 + x + x^2 + x^3 + \cdots = \sum_{n=0}^{\infty} x^n$$

It is critical to remember that, for the series to converge, the absolute value of x must be *less* than *1*, or as written in math terms, $|x|$ *must be* < 1

The importance of the geometric power series lies in the fact that we can substitute any *f(x)* for x (as long as *f(x)* remains less than *1*) into the geometric series, and obtain **power series expansions** that allow the *simple approximation* of the values of many mathematical functions.

As an example, let us substitute the function, $f(x) = -x^2$, for the original variable, q, rather than just plain x. The power series equation will *remain* valid, and we will have

$$\frac{1}{1 - (-x^2)} = \frac{1}{1 + x^2} = 1 + (-x^2) + (-x^2)^2 + (-x^2)^3 + \cdots$$

$$= 1 - x^2 + x^4 - x^6 + \cdots$$

We have generated a new function of x, $f(x) = 1/(1 + x^2)$, whose value can be approximated directly by the power series on the right of the equation. The power series can also be used to approximate the values of other functions of x such as

$$\frac{1}{1 + x} \quad (q = -x); \quad \frac{1}{1 - x^{-4}} \quad (q = x^{-4}); \quad \frac{3x^2}{1 + x} = 3x^2\left(\frac{1}{1 + x}\right) \quad or\ 3x^2(\frac{1}{1 + x}\ series)$$

Note that the function represented by the power series can be *just* a function, or it can be the *derivative* of a second function. In the latter case, the power series must be integrated to obtain the value of the desired second function.

The expression obtained above, $1/(1 + x^2)$, turns out to be the derivative of the **arctangent** function. We recall that the arctangent function, given the tangent of some angle, will return the value of that angle in degrees or radians, as for example,

$$\tan 31^0 = \tan 0.54105\ radians = 0.60086...$$

$$\arctan 0.60086... = 31^0 = 0.54105...\ radians$$

The derivative of the arctangent function is written as

$$\frac{d(\arctan\ x)}{dx} = \frac{1}{1 + x^2}$$

And if this derivative is integrated, we have

$$\arctan\ x = \int (1 - x^2 + x^4 - x^6 + \cdot\cdot\cdot)\ dx$$

$$= x - \frac{x^3}{3} + \frac{x^5}{5} - \frac{x^7}{7} + \cdot\cdot\cdot\cdot + C\ (\textit{integration constant})$$

Integrating each term in the power series separately (see Chap. 9 on calculus), we obtain the final integral (series) form for arctan x. The **integration constant, C**, must be evaluated. Let us set $x = 0$. The arctan of 0 is equal to 0, and the function of x, (i.e., $1/(1+x^2)$), reduces to 1 for
$x = 0$. The integral of $1dx$ equals x, and becomes 0 when $x = 0$. Therefore C is equal to 0.

We can test the approximation provided by this integral using the tangent and arctangent data from above. The tangent of the angle, x, was $0.60086...$, and using this value for x, we have

$$\textit{arctangent}\ x = 0.60086... - \frac{(0.60086...)^3}{3} + \frac{(0.60086...)^5}{5} - \frac{(0.60086...)^7}{7} + \cdot\cdot\cdot$$

$$= 0.60086... - 0.07231... + 0.01566... - 0.00404... = 0.54017...\ \textit{radian}$$

$$\textit{or in degrees,}\ \ 0.54017...\ \textit{radians} \times \frac{180^0}{\pi} = 30.949^0...$$

Notice that we use the value for the *tangent* in the various terms of the power series, and the value returned is the *angle* in radians. Multiplying the value in radians by $180^0/\pi$ then gives the angle in degrees. The precision of this approximation is about *5* parts in *3000*. If we include the 9th and the 11th power terms, our angle is indicated to be 31.006^0. More terms would given even more accuracy.

Fig. 8-1 shows the arctan x function plotted against two power

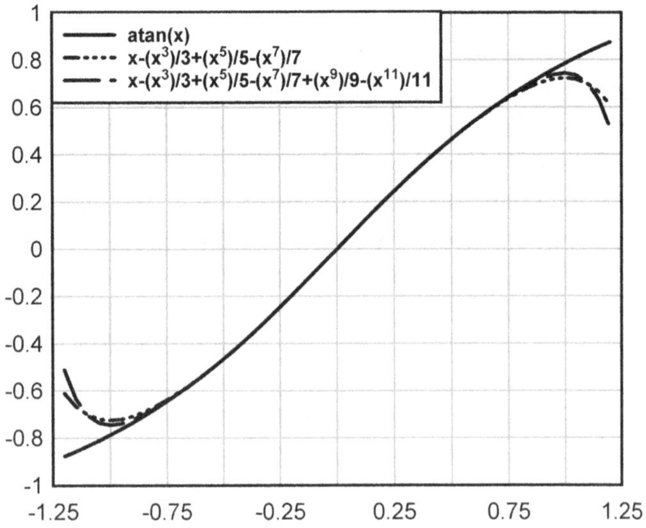

Fig. 8-1 Plot of arctan x vs. two power series in x

series in x, with one series going to 7th power and the other to the 11th power. Notice that the 11th power series is closer than the 7th power series to arctan x at *-1* and *1*. And also that once deviation begins, the 11th power curve turns away from arctan x *more* sharply than the 7th power curve.

This infers that a similar power series of *infinite* terms would come very close to (but never reach) arctan x at *-1* and *1*. And this is true as a consequence of the fact that the original geometric power series only *converges* when the values of q (or x) lie between $-1 < x < 1$. (This means that we can not use this method for angles greater than 45^0 (tan = *1*) or lesser than -45^0 (tan = *-1*).)

The interval of all the x's for which the power series converges is known as the **interval of convergence**. If the values of x fall within this interval, the power series will converge, and its use to appproximate a second function (such as the arctan function) will be "good." Mathematicians also speak of the **radius of convergence**. The determination of these two numbers involves finding the *limit* which the power series approaches as n (the number of individual terms) goes to infinity. The details of finding this limit are beyond the purpose of this tutorial, but the practical sense of convergence is discussed in many of the following examples.

The power series can also be used to calculate **logarithms**. The initial calculation involves what seems to be an unusual form — the $ln(1 - x)$. The reason for this is that $-1/(1 - x)$ is the derivative of $ln(1 - x)$, and if we multiply both sides of the geometric power series by *-1*, we obtain

$$-1 \times \frac{1}{1 - x} = -1 \times (1 + x + x^2 + x^3 + x^4 + \cdots)$$

And if we integrate this derivative series, we have the series expression for $ln(1 - x)$,

$$\ln(1 - x) = \int \frac{-1}{1 - x}\, dx = -\int (1 + x + x^2 + x^3 + x^4 + \cdots) dx$$

$$= -(x + \frac{x^2}{2} + \frac{x^3}{3} + \frac{x^4}{4} \cdots) + C = -\sum_{n=1}^{\infty} \frac{x^n}{n} + C$$

To evaluate the integration constant, C, we set $x = 0$. The left side of the equation is then $ln(1) = 0$, and all of the x terms on the right side $= 0$; therefore, $C = 0$.

Let us test this equation for $x = 0.3$. We are then calculating $ln(1 - 0.3)$ or $ln(0.7)$. My calculator shows $ln(0.7)$ to be *-0.356674943*, or -0.35667_5. We have then that

$$-0.35 [illegible] + [illegible] \frac{(0.3)^3}{3} [illegible] + \frac{(0.3)^5}{5} + \cdots)$$

$$[illegible] + [illegible] 0.009 [illegible] 0.0005 = 0.3565...)$$

And we see that aft[illegible] po[illegible] equati[illegible] mates the value of *ln(0.7)* to within *2* parts in *10*[illegible]

However, su[illegible] we wi[illegible] late [illegible] tly. To do this, we *substitute* the variable, *1 - x*, [illegible] e, *x*, i[illegible] iginal [illegible] ion. This gives

$$\ln(1-(1-x)) [illegible] = -[(1- [illegible])^2 [illegible] + \frac{(1-x)^4}{4} + \frac{(1-x)^5}{5} + ...]$$

[illegible] < [illegible]

And since we are s[illegible] *7*, eac[illegible] or b[illegible] .7) = 0.3. This series is just the same as the series al[illegible] *1-x)*, s[illegible] nd [illegible] swer for *ln(0.7)*.

Notice that [illegible] converg[illegible] valu[illegible] ween *2* and *0*. If *x* is greater than *2*, the value of [illegible] r in [illegible] will [illegible] r than *1*, and the series will not converge because e[illegible], is be[illegible] rger [illegible] rm, a_n, that preceded it. If *x* is less than *0*, then *x* i[illegible] umber [illegible] can[illegible] logarithm of a negative number.

Although thi[illegible] erges [illegible] we ca[illegible] etheless to approximate the logarithm of *any* nu[illegible] is pos[illegible] se

[illegible]

and for exam[illegible] *7* = ln [illegible] = [illegible] ln 2) + ln 1.3593...

However, there is a[illegible] ecaus[illegible] conv[illegible] logarithm of *2* (which is just at the limit of converg[illegible] lowly.

$$\ln(x) [illegible] + \frac{([illegible]}{} [illegible] \frac{(1-x)^3}{3} [illegible] + \frac{(1-x)^5}{5} + ...]$$

fo[illegible] = [illegible] $\frac{1^4}{4} + \frac{-1^5}{5} + ...)$

$$= -(-1 + \frac{1}{2} [illegible] - \frac{1}{5} [illegible]) = [illegible] (\textit{true } \ln(2) = 0.693147..)$$

After the 7th term, for example, the series has a value of *0.7595..*, while the true logarithm of *2* is *0.6931...*

We can determine how many terms would be required to give a good approximation for *ln(2)*, however, because the series is an *alternating* series. In an alternating series, the terms alternate between positive and negative, and each term is *less* than the one before it. The "error bound" (or limit of error) of an alternating series can be shown (pg. 147) to be equal to, or less than, the value of the term just after the truncation term. For example, we truncated here at the 7th term, and the value of the 8th term is *1/8* or *0.125*. The true value then lies between 0.759523 ± 0.125, or in other words, between *0.884523* and *0.634523*. But if we truncated at the 9999th term, the limit of error would be *1/10000*, and our *ln(2)* approximation would be good to $0.6931... \pm 0.0001$.

The calculation of *10,000* terms is easy for a computer, but difficult for humans. Early mathematicians therefore developed alternative series that converged more rapidly for the calculation of natural logarithms.

But note that numbers near the center of the convergence interval for this series do converge rapidly, as with the *ln(1.3593)*:

$$\ln(1.3593) = -(-0.3593 + \frac{(-0.3593)^2}{2} + \frac{(-0.3593)^3}{3} + \frac{(-.03593)^4}{4} + \frac{(-0.3593)^5}{5} + ...)$$

$$\approx 0.3073... \textit{ after 5 terms } (\textit{true } \ln(1.3593) = 0.3069...)$$

After only *5* terms, this series converges to within *4* parts in *10,000*. And if we complete the original calculation for *ln(87)* using this approximation, and the "9999th term" approximation for *ln(2)*, we obtain a good approximation for *ln(87)*, i.e.,

$$\textit{approx. } \ln(87) = 6(0.6931) + 0.3073 = 4.4659... \quad (\textit{true } \ln(87) = 4.465908119...)$$

The important mathematical fact here is that we often obtain good approximations, from the calculation of only a few terms, *only* if the value being approximated lies near the center region of the convergence interval.

Newton's Method for Solving Equations

After calculus became known, series formulae were developed that depended upon "..the *derivative* of the function"; that is, a mathematical concept unique to calculus. One of the first such calculations was Newton's Method for Solving Equations.

This method is often called a means for "solving for the zeros." In the section on plotting Third-Degree and Higher Polynomials (pg. 46), we saw that if we plot a polynomial such as $x^3 - 6x^2 + 11x - 6$, the points at which the curve intersects the x axis (i.e., where the polynomial equals *0*) represent the *solutions* to the polynomial equation. The solutions are also known as the **factors** of the equation (factors are the *terms*, which when multiplied, produce the original equation). This particular polynomial curve intersects the x axis at the points *1*, *2*, and *3*, and the factors of the equation are thus $x - 1$, $x - 2$ and $x - 3$. Multiplying these three factors regenerates the original equation.

There are two difficulties with solving equations by plotting. Plotting is laborious (before computers), and the intersection of the curve with the x axis can not be read with *precision*. Newton's method avoids these problems.

Newton's method is based on the fact that the *first derivative* of a function, and let us write the function as $f(x_n)$ and its first derivative as $f'(x_n)$, produces a tangent line whose slope is

$$f'(x_n) = \frac{\Delta y}{\Delta x} = \frac{f(x_n) - 0}{x_n - x_{n+1}}$$

This is illustrated in Fig. 8-2. Notice that the value of y on the curve is simply equal to $f(x_n)$. And that we can rearrange the preceding equation to obtain

$$x_{n+1} = x_n - \frac{f(x_n)}{f'(x_n)}$$

The advantage of this last equation is that we do not have to actually draw the graph; rather we can now work simply from the function.

Let us find the square root of *18*. In equation form, this becomes

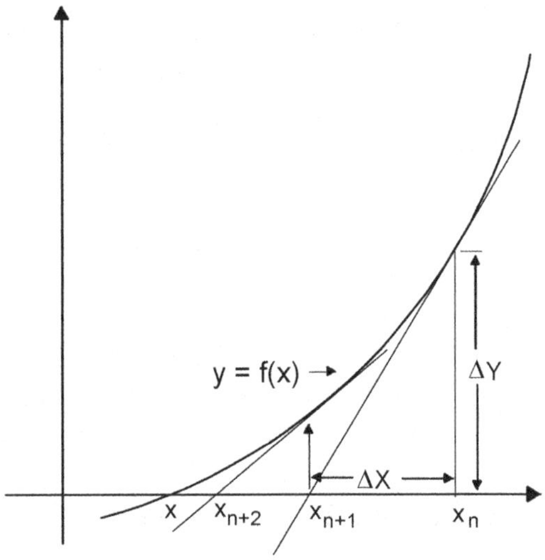

Fig. 8-2 Iteration by Newton's method

$$f(x_n) = x_n^2 - 18 = 0$$

$$\textit{and the first derivative, } f'(x_n), = 2x_n$$

The first derivative is obtained from calculus by the rule of differentiation for x^n (note that the derivative of the constant, *-18*, is zero) which is

$$\frac{d\,[f(x)]}{dx} = \frac{d[x^n]}{dx} = nx^{n-1}$$

We know that the square root of *18* falls between *4* (square root of *16*) and *5* (square root of *25*), and in fact, our calculator gives the square root of *18* as *4.242640687*. We make a first guess of *4.5*, which corresponds to x_n in Fig. 8-2 (which is not to scale), and so

$$x_{n+1} = x_n - \frac{f(n)}{f'(n)} = 4.5 - \frac{(4.5)^2 - 18}{2(4.5)} = 4.25$$

We then do a second iteration using the new calculated value for x_{n+1}, i.e.,

$$x_{n+2} = 4.25 - \frac{(4.25)^2 - 18}{2(4.25)} = 4.242647059$$

This new value agrees out to five decimal places with the value for the square root of *18*. As a matter of interest, let us do a third iteration using the x_{n+2} value of *2.242647059*,

$$x_{n+3} = 4.242647059 - \frac{(4.242647059)^2 - 18}{2(4.242647059)} = 4.242640687$$

This last answer agrees completely with the value for the square root of *18* given by our calculator, and if we square this answer, our calculator shows an *integer* of *18*; that is, the error is so small that the calculator "rounds off" to an answer of *18*. Answers of high precision can be obtained therefore from Newton's method.

As another example, let us consider the third-degree polynomial sketched in Fig. 8-3. As we have seen on pgs. 44-47, a curve of this form has only one *real* answer (i.e., an answer in real numbers) which appears to be at about *0.5*.

If we did not have the sketch, we could find the approximate solution by a trial-and-error method, where we set arbitrary values for *x* and calculate the resultant *y*. We see that *y* goes from *1* to *-0.496* when *x* goes from *1* to *0.4*, so *y = 0* must lie between *x = 1* and x = 0.4.

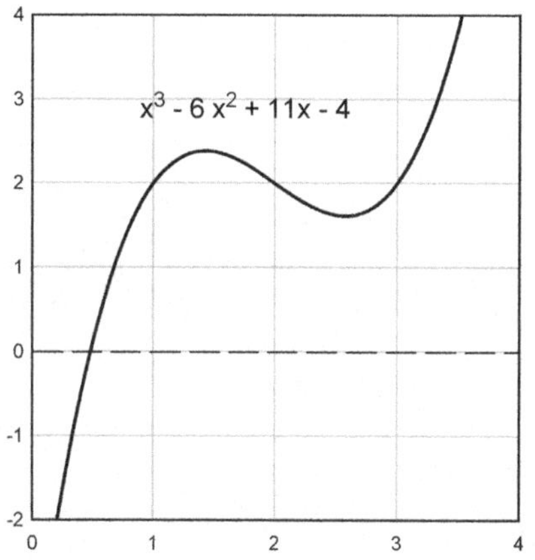

Fig. 8-3 Graph of third-degree polynomial

Trial-and-Error Calculations

f(n)	Assumed x	Resultant y
$x^3 - 6x^2 + 11x - 4$	4	8
(*same*)	3	2
(*same*)	2	2
(*same*)	1	2
(*same*)	0.4	-0.496

We will star[illegible]r's met[illegible]ions [illegible]he guess that $x_n = 0.4$, and knowing that

$$[illegible] - 6x [illegible] 4$$

$$[illegible] and, f'[illegible] - 1[illegible]$$

Our first calculation [illegible]

$$x_{n+1} [illegible] \frac{(.4)^3 - [illegible]}{3([illegible] [illegible](0.4)} = 0.474251497$$

and then a second i[illegible]

$$x_{n+2} = 0.4742[illegible] - \frac{(0.4742[illegible] - 6([illegible])^2 + 11(0.474251497) - 4}{[illegible] [illegible]149[illegible] [illegible].474251497) + 11}$$

$$[illegible] 786[illegible]$$

And substituting thi[illegible] x_{n+2} int[illegible]iginal eq[illegible] find that

$$y = (x_{n}[illegible]^2 + 11[illegible] - 4 = [illegible] \quad \textit{or essentially zero}$$

Newton's m[illegible]ore sh[illegible]e r[illegible]f the polynomial equation to a precision of at least [illegible].

However, N[illegible]thod [illegible]always [illegible]ll as with these two examples. If the starting valu[illegible] ch[illegible]cces[illegible]mations may not converge, and will lead away fro[illegible] answ[illegible]ith [illegible]tions, the rate of convergence will be much slowe[illegible] One [illegible]alyz[illegible]ns under study carefully before applying Newton's [illegible]

Maclauri[illegible]lor Se[illegible]

While New[illegible] uses [illegible]st de[illegible] the function, the Maclaurin and Taylor series use th[illegible]ive, 2nd [illegible], 3rd [illegible] 4th *derivative, and so on* .. of the function. We th[illegible] func[illegible]e continually differentiated. The three functions [illegible]non[illegible]st-[illegible]s are

$$e^x; \quad where \quad \frac{d(e^x)}{dx} = e^x$$

$$\sin x; \quad where \quad \frac{d(\sin x)}{dx} = \cos x$$

$$\cos x; \quad where \quad \frac{d(\cos x)}{dx} = -\sin x$$

The Maclaurin series (variable = x) is just the Taylor series (variable = x - a) for the special case where $a = 0$. The two series are separately named mostly for historical reasons.

The basis behind the Maclaurin (or Taylor) series treatment derives from polynomials. Consider a third-degree polynomial function, $f(n)$, which we write in "reverse" form,

$$f(n) = y = 6 + 3x^1 + 4x^2 + 2x^3$$

$$or\ in\ general\ form, \quad y = a_0 + a_1x^1 + a_2x^2 + a_3x^3$$

The numbers, a_n, are the coefficients of the terms containing the increasing powers of x. This polynomial generates a curve where, for *every* x, there is a single, unique value of y.

We can reproduce the values of the coefficients by analyzing the curve at the point on the curve where $x = 0$. We look first at the value of the original polynomial for $x = 0$, and then at the values of the *1*st, *2*nd, and *3*rd derivatives of the polynomial for $x = 0$. This gives

$$f(n) = 6 + 3x + 4x^2 + 2x^3; \qquad a_0 = 6$$

$$f^{I}(n) = 3 + 8x + 6x^2; \qquad 1 \cdot a_1 = 3, \ a_1 = \frac{3}{1!} = 3$$

$$f^{II}(n) = 8 + 12x; \qquad 1 \cdot 2 \cdot a_2 = 8, \ a_2 = \frac{8}{2!} = 4$$

$$f^{III}(n) = 12; \qquad 1 \cdot 2 \cdot 3 \cdot a_3 = 12, \ a_3 = \frac{12}{3!} = 2$$

Note that in differentiating the various terms of the polynomial, we are simply following the rule for the differentiation of x^n, and that is where the *factorial divisors* come from, i.e.,

$$\frac{d(x^n)}{dx} = nx^{n-1};\ \ for\ a_0x^n,\ \ \frac{d(a_0x^n)}{dx} = a_0nx^{n-1}$$

$$1^{st}\ derivative\ of\ a_3x^3 = 3a_3x^2$$
$$2^{nd}\ derivative\ of\ 3a_3x^2 = 2\cdot 3a_3x^1$$
$$3^{rd}\ derivative\ of\ 2\cdot 3a_3x^1 = 1\cdot 2\cdot 3a_3$$

$$so,\ if\ 3^{rd}\ derivative\ of\ a_3x^3\ term = 12\ (as\ above),\ then\ a_3 = \frac{12}{3!} = 2$$

The **Maclaurin series** is based on an extension of this idea; that is, if a given function can be represented by a power series, then the *coefficients* of that series are given by

$$a_n = \frac{f^{(n)}(0)}{n!}$$

Beginning students often fail to recognize the importance of coefficients in a power series. It is in fact the coefficients which determine the *value* of the power series. Also, if one determines the equation that produces a curve from data at *one* point on the curve, that *same* equation generates *all* of the points on the curve.

The Maclaurin series is written in mathematical form as

$$f(x) = \sum_{n=0}^{\infty}\frac{f^{(n)}(0)}{n!}x^n = f(0) + f'(0)\cdot x + \frac{f''(0)}{2!}\cdot x^2 + \frac{f'''(0)}{3!}\cdot x^3 + \ldots + \frac{f^{(n)}(0)}{n!} + \ldots$$

In this equation, *f(x)* is a function that can be represented, or we could say, reproduced by some power series. The summation expression,

$$\sum_{n=0}^{\infty}\frac{f^{(n)}(0)}{n!}x^n$$

says that *f(n)* can be reproduced by summing, from $n = 0$ to $n = \infty$, the 0st, 1st, 2nd, 3rd, ... up to the nth derivatives of *f(n)*, where each derivative is divided by the factorial of the number of the derivative. The value of each *coefficient* is determined at the point, $x = 0$, and this coefficient is then multiplied by x^n.

The "instructions" given by the summation expression can be more easily seen if we look at the series expansion at the right of the equation,

$$f(0) + f'(0)\cdot x + \frac{f''(0)}{2!}\cdot x^2 + \frac{f'''(0)}{3!}\cdot x^3 + \ldots + \frac{f^{(n)}(0)}{n!} + \ldots$$

And this expansion series itself can be more easily understood simply by doing an example calculation such as $f(x) = e^x$:

$$f(x)=e^x \textit{ and } f(x) = f(0) = e^0 = 1, \textit{ so } a_0 = 1$$

$$\underline{\textit{and}}\ f'(x), f''(x), f'''(x) \ \ldots \textit{ all } = e^x, \textit{ and thus } \underline{\textit{all}} = 1 \textit{ for } x = 0$$

$$\textit{therefore the } \underline{\textit{numerator}} \textit{ of all coefficients } (a_0 \rightarrow a_\infty) = 1$$

And when the *numerator* of all of the term coefficients is equal to *1*, the series expansion becomes

$$e^x = 1 + x + \frac{x^2}{2!} + \frac{x^3}{3!} + \frac{x^4}{4!} + \frac{x^5}{5!} + \ldots = \sum_{n=0}^{\infty} \frac{x^n}{n!}$$

This is the Maclaurin series for e^x. It can be shown that this series converges for all values of x between $-\infty$ *and* ∞. This series can be used therefore to calculate the value of e^x for all real numbers. For example, let $x = 2$.

$$e^2 = 1 + 2 + \frac{2^2}{2!} + \frac{2^3}{3!} + \frac{2^4}{4!} + \frac{2^5}{5!} + \frac{2^6}{6!} + \frac{2^7}{7!} + \ldots$$

$$= 1 + 2 + 2 + 1.333 + 0.667 + 0.267 + 0.089 + 0.025 + \ldots = 7.381 + \ldots$$

Our calculator gives $e^2 = 7.38906...$ The series does not converge rapidly, but the approximation is reasonably good. More terms would make the approximation better. One can also turn the argument around, and say that the ln of *7.381* is approximately *2* (actually *1.9989...*).

As another example, let us determine the Maclaurin series for *cos(x)*. Since *cos(0) = 1* and *sin(0) = 0*, and using the derivatives for *sin* and *cos* given above, we find that the derivatives of *cos(x)* at $x = 0$ are [illegible] *of* [illegible] *Val*[illegible] *ative at x*=0

[illegible] (0))
[illegible] in(0)) [illegible]
[illegible] os(0) [illegible]
[illegible] (0))
[illegible] (0))

. . .

If we use th [illegible] the [illegible] Maclaurin series expansion, we find that the value [illegible] proxi [illegible]

$$[illegible] = 1 [illegible] ...$$

As a test, let us find [illegible] of *0.5* [illegible] which [illegible] ulator is 0.87758...
Substituting *0.5* for [illegible]

$$[illegible] = 1 [illegible] + ...$$

$$[illegible] 125 + [illegible] 0.87758...$$

The Maclaurin seri [illegible] us c [illegible] agreement through five decimal places with [illegible] or a [illegible]
Early mathe [illegible] uld n [illegible] those given by a calculator.
The goodness of th [illegible] nation [illegible] interest. For an **alternating series**, where the te [illegible] betw [illegible] *and* each term is less than the previous one, they [illegible] how th [illegible] was equal to, or less than, the next term in the seri [illegible] in th [illegible] *cos(0.5)* , for example, would be equal to, or less [illegible]

$$[illegible] or \leq [illegible] 0.000000969$$

The approximation [illegible] he M [illegible] or the cosine function is therefore quite good [illegible]

The determination of the limit of error, or "error bound", for an alternating series has been much studied. This is one way of looking at the problem. The series equation representing *cos x* is

$$\cos x = S = \sum_{n=0}^{\infty}(-1)^n\frac{x^{2n}}{2n!} = 1 - \frac{x^2}{2!} + \frac{x^4}{4!} - \frac{x^6}{6!} + \frac{x^8}{8!} - \frac{x^{10}}{10!} + \ldots$$

In this equation, S is the **total sum** of the series when *n* goes to infinity. It is the true answer for *cos x*. We can find an approximate answer, however, by taking the **partial sum,** S_n, which is the sum of the terms up to some arbitrary stopping term. In approximating *cos 0.5*, for example, we stopped the summation at the fourth term. We label this partial sum as S_3 because the value of *n* is *3* even though it is in the fourth term. (Under this labeling scheme, the first partial sum would be S_0, because *n* in the first term is equal to *0*).

The taking of a partial sum effectively splits the original summation into two parts, as in this case

$$S = \sum_{n=0}^{\infty}(-1)^n\frac{x^{2n}}{2n!} = \sum_{n=0}^{3}(-1)^n\frac{x^{2n}}{2n!} + (\textit{the truncation error}) \sum_{n=4}^{\infty}(-1)^n\frac{x^{2n}}{2n!}$$

For the error bound approximation to be true, we need to show that the value of the fifth term, which we will call a_5, is greater than the partial sum, S_4, or

$$S_4 = \sum_{n=4}^{\infty}(-1)^n\frac{x^{2n}}{2n!} \leq |a_5|$$

In English, we are saying that the partial sum S_4, which is the partial sum of the alternating series from the fifth term a_5 down to the infinity term, a_∞, (and which *includes* the fifth term itself), is less than the absolute value of the fifth term. The absolute value of the term is used because the term in question could be positive or negative.

This is possible because 1) each term is smaller than the preceding term, and 2) the terms alternate between positive and negative. A common illustration to explain the error bound behavior of an alternating series is shown in Fig. 8-4. The *values* of the terms of the series are shown as *a(n)*, *a(n+1)*, *a(n+2)*, ...,

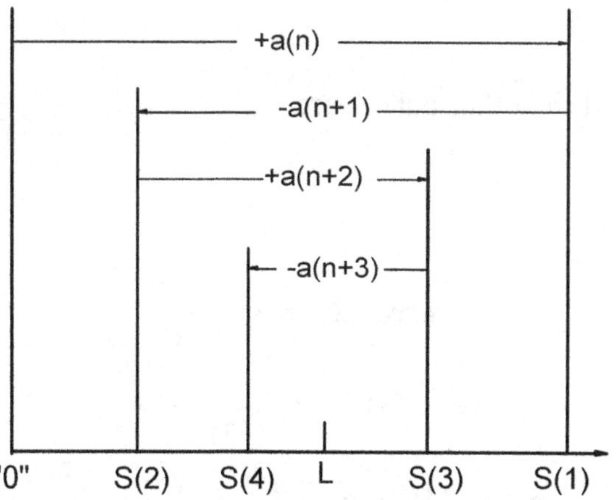

Fig. 8-4 Error bound for alternating series

and the "residual" sum (i.e., the sum for all of the remaining terms to $n=\infty$) associated with the term is given as *S(1)*, *S(2)*, ... The limit, L, is the sum of the series at the *infinity* term, a_∞.

In Fig. 8-4, the sum of *+a(n)* and the following *-a(n+1)* is equal to the distance between *"0"* and *S(2)* on the horizontal axis. The sum of *-(n+1)* plus *+a(n+2)* equals the distance between *S(3)* and *S(1)*. The sum *+a(n+2)* plus *-a(n+3)* equals the distance from *S(2)* to *S(4)*, and so on. Most importantly, note that the sum of *all* distances on the horizontal axis can not be *greater* than *+a(n)*, i.e., the distance from *"0"* to *S(1)*. This is the error bound theorem.

Maclaurin vs Taylor series. The only difference is that the variable in the Maclaurin series is "*x*", whereas in the Taylor series it is "*x -a* ". And the *only* effect of the constant, *a*, is to shift the function curve along the *x* axis, as shown for the polynomials in Fig. 8-5. The original curve has been moved *4* units to the right by changing the variable to "*x - 4*".

One might ask, why would anyone want to do this? What is the advantage of the Taylor series? The answer is that series approximations made for points near "zero" are more precise, *and* require fewer terms, than those for points far from zero. This is shown in Figs. 8-6 and 8-7.

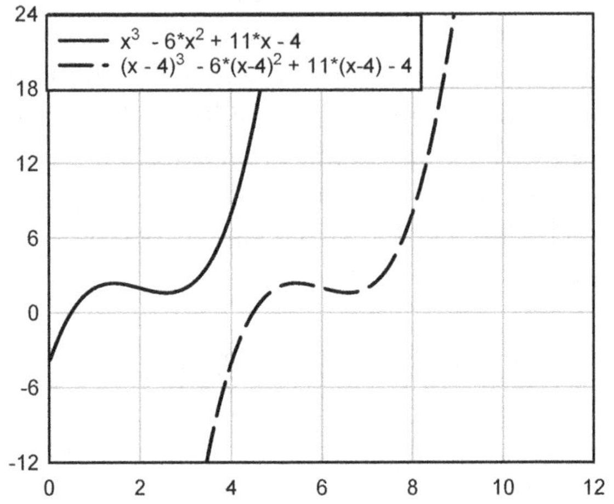

Fig. 8-5 Changing x to (x -4) only moves curve

In Fig. 8-6, we use the Maclaurin series to approximate $y = sin(x)$ when $x = 1.5\pi$. The Maclaurin series for *sin(x)* is

$$\sin(x) = x - \frac{x^3}{3!} + \frac{x^5}{5!} - \frac{x^7}{7!} + \ldots$$

Curve (1) in Fig. 8-6 is the true sine curve.

And the other curves are

$$curve\ (2) = x - \frac{x^3}{3!}$$

$$curve\ (3) = x - \frac{x^3}{3!} + \frac{x^5}{5!}$$

$$curve\ (4) = x - \frac{x^3}{3!} + \frac{x^5}{5!} - \frac{x^7}{7!}$$

$$curve\ (5) = x - \frac{x^3}{3!} + \frac{x^5}{5!} - \frac{x^7}{7!} + \frac{x^9}{9!}$$

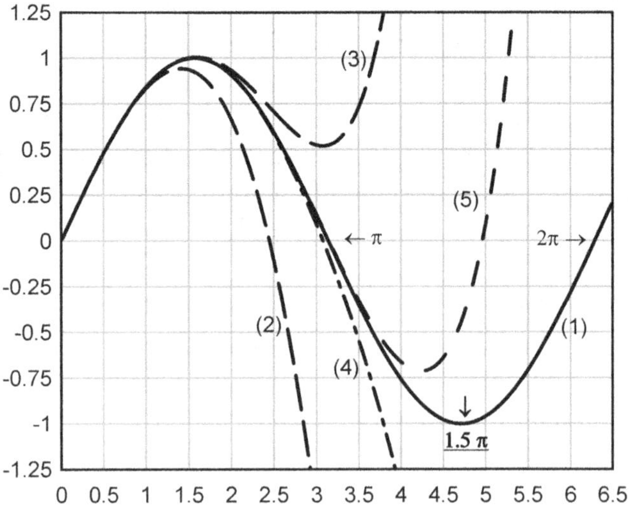

Fig. 8-6 Maclaurin series approximation for x = 1.5π

Notice that the true value of *sin(1.5π)* is *-1*, but even curve (5) indicates the value to be about *-0.25* — a very poor approximation.

In contrast, the Taylor series approximation in Fig. 8-7 for the value of *sin(x)* at $x = 1.5\pi$ is very good between the values of $x = 4$ to $x = 5.5$ *radians* even for curve (2). Clearly one would rather use the Taylor series in this case where we are seeking the value of *sin(x)* at a point well removed from $x = 0$.

The Taylor series approximation for *sin(x)* centered at 1.5π, from which curves (2) through (4) are derived, is given by

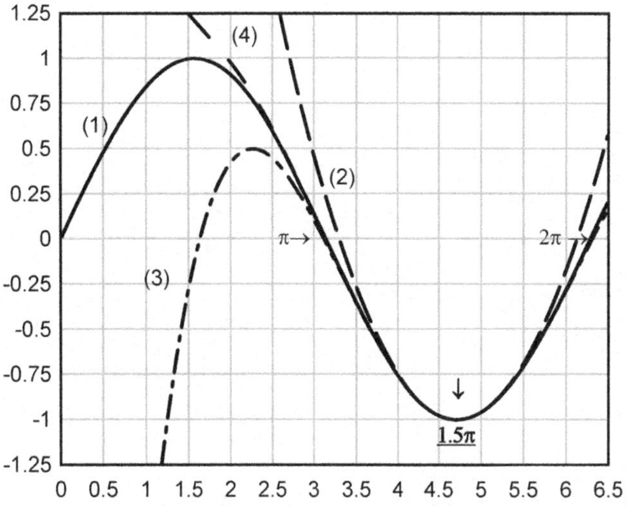

Fig. 8-7 Taylor series approximation for x = 1.5π

$$\sin(x) = -1 + \frac{(x-1.5\pi)^2}{2!} - \frac{(x-1.5\pi)^4}{4!} + \frac{(x-1.5\pi)^6}{6!} + \ldots$$

$$\textit{with curve } (2) = -1 + \frac{(x-1.5\pi)^2}{2!}$$

and curves (3) *and* (4) *including the* 3rd *and* 4th *terms, resp.*

If *a* consists of different multiples of π, we find that the Taylor series for *sin(x-a)* always produces at least a certain region wherein the Taylor approximation gives a good value for *sin x*, as shown in Fig 8-8. And notice that this region is *centered* around the given value for π, and that the region becomes larger as more terms are included in the series.

But note further, however, that the "*sin(x)*" equation used to plot the Taylor series approximation for $x - 1.5\pi$ is *not* simply the Maclaurin series for *sin(x)* with *x* replaced by $x - 1.5\pi$. The three relevant functions are given below, where we see that the Taylor series for

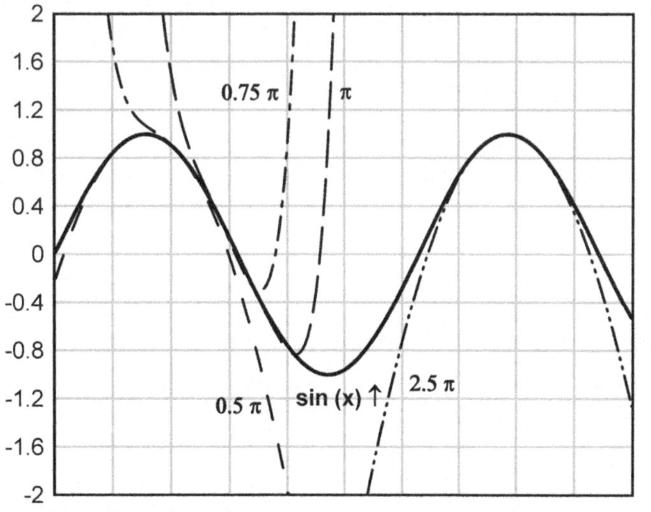

Fig 8-8 Regions of convergence for different multiples of π

sin x is actually the ... the M... ...ie... ...ith *x-1.5π* being substituted for the *x* variable:

... *Maclaurin*)

$$... 1 - \frac{x^2}{2!} ... \frac{x^6}{6!} + ... \quad (Maclaurin)$$

$$\sin x ... \frac{(x-1.5\pi)^2}{2!} ... (x-1.5\pi)^4 ...)^6 + ... \quad (Taylor)$$

We explore ... this ... Fig. 8-9, where *cu*... ...), *cu*... = *cos(x)*, and *curve*... ...l from ... same equation that ... ...rve (4) ... plotting of the Tay... ...*sin(*... Fig. 8-7.

Curves (4) ... ...g 8-9 ... plotted from the M... ...es for ... where we have sim... ...he *x* variable by *x-1.5π* ... ...iables respectively. Each ... ...ions is limited to just three ...

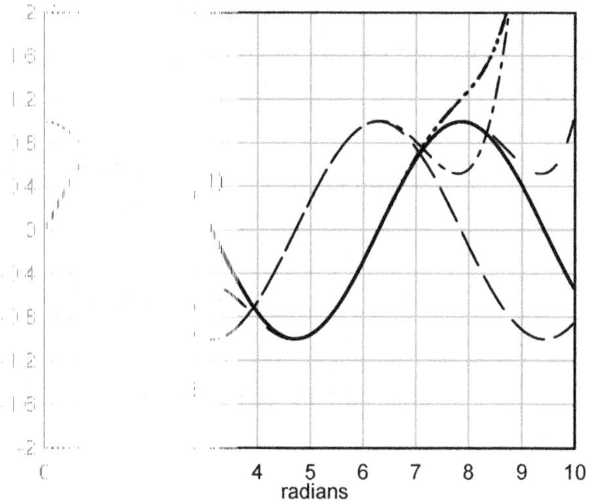

Fig 8-9 Taylor series for cos (x - a) and sin (x - a)

$$...rve\ (4) ... \pi) ... + \frac{(x-1.5\pi)^5}{5!}$$

$$...rve\ (5) ... \frac{(x-2\pi)^5}{5!}$$

In Fig. 8-9, ... curve ... too f... ...t, and in fact coincides well with the *cos(x)* cur... ...2) b... ...ns. But we are seeking an approximation for ... ...se t... ...ab... curve (5) which is moved *0.5π* to the right, and wh... ...proxim... ...ver th... about *4* to *7* radians, and especially well at *x*...

This is puzzl... ...sight, bu... ...rom the trigonometric identity,

$$\sin(x - 0.5\pi) = -\cos(x)$$

$$or,\ \sin((x-a) - 0.5\pi) = -\cos(x-a)$$

If $a = 1.5\pi$, then we have

$$\sin(x - 2\pi) = -\cos(x - 1.5\pi)$$

The negative of the cosine Taylor series, $-cos(x - 1.5\pi)$, is equal therefore to the sine Taylor series, $sin(x - 2\pi)$, and gives us the desired approximation for the sine curve centered around $x = 1.5\pi$.

Goodness of Fit

Early-day mathematicians did not have the luxury of computer plotting. They had to rely on mathematical reasoning to evaluate the "goodness" of the fit between the Taylor series and the function to be approximated. One method was developed by **J. Lagrange** (*1736-1813*). He proposed that, if the function has $n+1$ derivatives in a given interval that contains both a and x, then there is a number **z** which occurs *between* a and x such that

$$f(x) = S_n(x) + Remainder\ R_n(x) \left(= \frac{f^{(n+1)}(z)}{(n+1)!}(x-a)^{n+1} \right)$$

The term, $f(x)$, represents the total value of the series at $n=\infty$, $S_n(x)$ is the partial sum of the series up to the nth term, and $R_n(x)$ is the value of the remaining terms as calculated from the value of the $(n+1)^{th}$ derivative of $f(x)$ at x = **z**, and where this value is then multiplied by $(x - a)^{n+1} / (n+1)!$

The derivation of this relationship is too sophisticated for the scope of this tutorial, but we can demonstrate its use by evaluating the *possible error* (or "error bound") in the Maclaurin series ($a=0$) for $sin(x)$ as plotted in Fig. 8-6 according to the function,

$$\sin(x) = x - \frac{x^3}{3!} + \frac{x^5}{5!} - \frac{x^7}{7!} + \ldots$$

To obtain the error bound, the Lagrange equation is modified to take into account the fact that the absolute value of the term $|f^{(n+1)}(\mathbf{z})|$ is equal to, or less than, the **maximum term, M**, which occurs for all **z** between x and a. This gives

$$error\ bound = |f(x) - S_n(x)| = |R_n(x)|\ which\ is \leq M \cdot \frac{|x-a|^{n+1}}{(n+1)!}$$

For $f(x) = sin(x)$, each $|f^{(n+1)}|$ is either $|sin(x)|$ or $|cos(x)|$, for which the maximum value is *1*, so ***M* = *1***. We will take the intervals of *x* to be studied to be from *0* to π/4 (*0.79* radians), *0* to π/2 (*1.57* radians), *0* to π (*3.14* radians), and *0* to *1.5π (4.71* radians), and *n* to be *4*. For the first interval, the error bound is

$$error\ bound \leq 1 \cdot \frac{(\frac{\pi}{4})^5}{5!} \approx 0.0025$$

Using the same calculation, we find the error bound of the interval up to *π/2* to be ~*0.08*; the error bound of the interval up to *π* to be ~*2.5*; and the error bound of the interval up to *1.5π* to be ~*19*. Comparing these results with Fig. 8-6, we see that the error bound is very small for the regions in which the *sin(x)* curve and Taylor series approximation curves coincide (which is consistent), but then becomes grossly out of range when the curves begin not to coincide, particularly as at *1.5π*. The La Grange error bound therefore tells us essentially what we see in Fig. 8-6, but by an easy calculation and without the necessity of plotting.

This is a simple demonstration of error bound calculations for Taylor series polynomials (see also the error bound for alternating series, pg. 147), but it shows the reasoning involved in such calculations.

Chapter 9.

Basic Calculus

The basic concepts of differentiation and integration were introduced on pgs. 25-29. We will take up the study again by looking at a copy of Fig. 9-10 where the *curve* is determined by the **function**, $f(x)$; the *line* is the instantaneous slope of the curve as given by the **derivative**, $f'(x)$; and the *hatched region* is the area under the curve between $x(2)$ and $x(1)$ which can be calculated from the **integral**, $F(x)$.

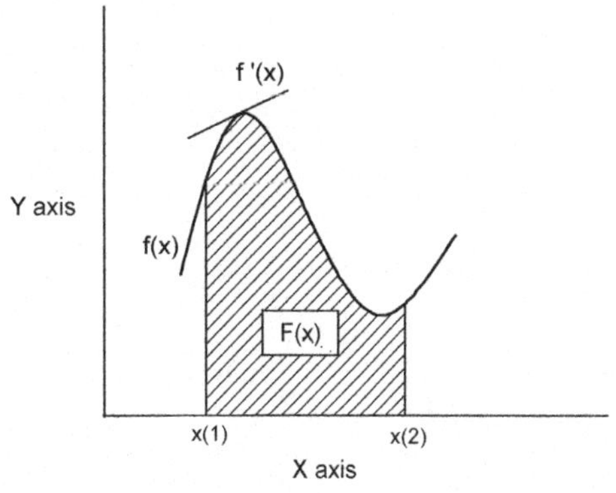

Fig. 9-10 Relation of function, derivative and integral

Starting from basics, we see that calculus is possible "...because something (the dependent variable) is a *function* of something else (the independent variable(s))." This idea is usually introduced as

$$y = f(x)$$

where y is the dependent variable (or "unknown"), and $f(x)$ signifies that y is a function of x. The important thing about a function — indeed, the *critical* thing — is that for every value of x there is a corresponding, and only *one*, value for y. Therefore, if x changes in value, then y must change by some certain fixed amount. The *amount* of this change is determined by the *equation* that constitutes the function, $f(x)$.

Calculus consists of just two basic operations; differentiation and integration. We will look at differentiation first.

Differentiation

The terms *differentiate*, *differential*, and *derivative* can confuse beginning students. In this tutorial, we use the term, *differentiate*, to describe what we do to find the difference in y that results from a change in x. What we find, or in other words, what *derives* from these operations, we call the *derivative*. The expression, dy/dx, is often referred to as the *differential* (as in differential equations), and we see in calculus equations that *dy/dx (the differential) = f'(x) (the derivative)*. It is important to keep in mind how the differential and derivative are related.

Math handbooks provide the student with tables of derivatives (also called differentials in older books) which are given in forms such as these

[illegible]

We will look at a n[illegible] more [illegible] used [illegible]s and show how they were derived, and also h[illegible] to th[illegible] fun[illegible]

One can rem[illegible] definit[illegible] ifferent[illegible] ting that differentiation shows the *difference* that [illegible] hen x [illegible].

The functio[illegible] ase ge[illegible] ong[illegible] rding to

(cons[illegible]
a line [illegible] *A)*
B ([illegible] *by B)*
a pow[illegible]
( ), cos [illegible] igono[illegible] *ession)*
and, [illegible] *or e* [illegible] *mic* [illegible] *tial expression)*

We will look at the [illegible] n of [illegible] se f[illegible] turn:

f(x) = C. S[illegible] nstant, [illegible] ch[illegible] Therefore the change in C with respect to x is

[illegible] = 0

f(x) = Ax. [illegible] uation [illegible] ght li[illegible] passes through *0,0*, with a slope of A. A is also the [illegible] the [illegible] he s[illegible] aight line is equal to

$$\frac{y_2 \ldots}{x_2 \ldots} \quad [illegible] \quad A\cdot\Delta x$$

A "Δ chang[illegible] s a lar[illegible] ist, [illegible] nge in y and x. But because the slope of a straight li[illegible] and do[illegible] pend [illegible] th of the segment, we could equally well make a very small change in x $(dx << 1)$, and we would find that $dy/dx = A$. Or to restate,

$$(large\ change)\ \frac{\Delta y}{\Delta x} = A; \quad (very\ small\ change)\ \frac{dy}{dx} = A$$

$$and\ since\ y = Ax,\ \frac{dy}{dx} = \frac{d(Ax)}{dx} = A, \quad and\ \mathbf{dy} = \mathbf{Adx}$$

In English this says that if x changes by dx, then y will change by $A \times dx$. And more importantly from the standpoint of calculus, if we make a number (say k) of small changes in x, the *summation* of these changes will be equal to the summation of the changes in y, or

$$A \sum_{n=1}^{k} dx = \sum_{n=1}^{k} dy$$

Note that the slope, A, can be equal to 1, and then we would have just $dy = dx$. As a "numbers" example, let the slope, A, be equal to 2, k equal 10, and $dx = 0.1$, then we have

$$\sum_{n=1}^{10} dy = \Delta y = 2 \cdot \sum_{n=1}^{10} (dx = 0.1) = 2$$

f(x) = Ax + B. The third modification that we can make to a linear equation is to add a constant, i.e.,

$$y = Ax + B$$

$$or\ y = f(x),\ where\ f(x) = Ax + B$$

The constant B is *not* a function of x. Changing x does not affect its value. Therefore, the differential of *any* constant (i.e., its change with respect to x) is simply equal to 0. The effect of the constant B is only to move the line, $y = Ax$, up or down the y axis. So we see that

$$dy = d(Ax + B) = d(Ax) + d(B) = A \cdot dx + 0, \quad so\ \mathbf{dy} = \mathbf{d(Ax + B)} = \mathbf{A \cdot dx}$$

Note that $dy = Adx$ gives the *change* in y that results from a change, dx, in x. To know the absolute value of y after this change, however, we must know B.

f(x) = x ⁿ. This is a **power term**, which is simply x raised to some power n, or in other words, x^n. We have shown (pg. 27) how to differentiate x^2 in terms of the derivative being the slope to the curve in Fig. 2-1, where that curve is generated by $f(x) = y = f(x^2)$. One adds a vanishingly small value, which we term h, to x to give $x + h$, and then evaluates the difference between $f(x + h)$ and $f(x)$ as $h \to 0$. As $h \to 0$, the slope of the secant → slope of the tangent, and we have,

$$\textit{slope of secant} = \frac{y_2 - y_1}{x_2 - x_1} = \frac{f(x_2) - f(x_1)}{x_2 - x_1} = \frac{f(x + h) - f(x)}{(x + h) - x} = \frac{f(x + h) - f(x)}{h}$$

$$\textit{since } f(x) = x^2\textit{, we then have } \frac{(x + h)^2 - x^2}{h} = \frac{(x^2 + 2xh + h^2) - x^2}{h} = \frac{h(2x + h)}{h}$$

$$\textit{slope of secant (or tangent)} = 2x + h \to 2x\textit{, as } h \to 0$$

$$\textit{slope of tangent} = 2x = f'(x) = (\textit{the } \underline{\textit{derivative}})$$

We can use the same approach to determine the derivative of *any* power term. For example, if $f(x) = x^4$, then

$$\frac{(x+h)^4 - x^4}{h} = \frac{(x^4 + 4x^3h + 6x^2h^2 + 4xh^3 + h^4) - x^4}{h} = 4x^3 = f'(x^4)$$

Therefore $f'(x^4)$, the first derivative of x^4, is equal to $4x^3$. And if we look at the form of the preceding equation, we see that as $h \to 0$, the general form of the derivative of x^n will always be

$$d\frac{(x^n)}{dx} = nx^{n-1}$$

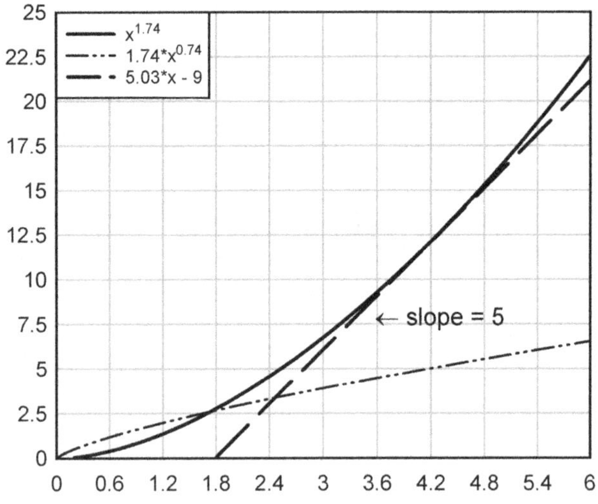

Fig. 9-1 Relationship of derivative to slope of curve

The use of n to express the exponent can cause the student to think that n must be an integer and positive. This is not true. In Fig. 9-1, we have plotted $y = f(x) = x^{1.74}$, and its derivative, $f'(x) = 1.74\,x^{0.74}$. The derivative curve gives the *slope* of the function curve at *every* value of x. For example, at $x = 4.2$, the value of the derivative curve is

5.03. If we draw a line with a slope of 5 to the original function curve at $x = 4.2$, we see that it is indeed the tangent line to the function curve at this particular value of x.

In Fig. 9-2, we examine the case where the power is *negative* by plotting $y = x^{-3}$ and its derivative, $f'(x^{-3}) = -3\,x^{-4}$. The curve for x^{-3} is discontinuous, and forms one asymptotic curve in the upper right-hand quadrant of the graph, and a second in the lower left-hand quadrant. The curve for $-3x^{-4}$ is also discontinuous, and forms asymptotic segments in the lower left-hand and lower right-hand quadrants of the graph.

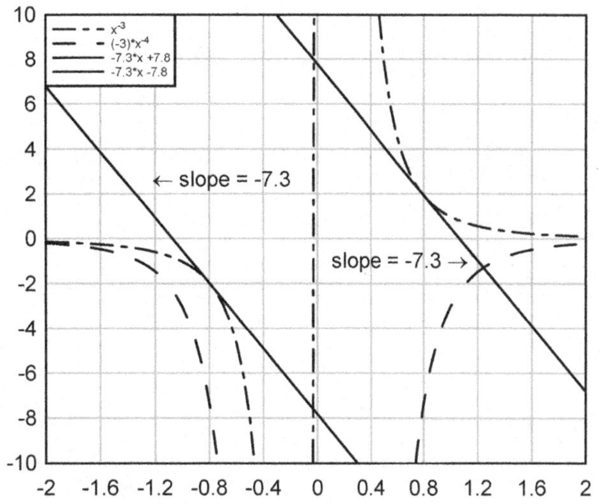

Fig. 9-2 Curves with negative exponent of x, and its derivatives

The important point, however, is that the value of the derivative curve is -0.73 at both $x = -0.8$ and $x = 0.8$. And if we draw a line with a slope of -0.73 to the original x^{-3} asymptotic curves at $x = -0.8$ and $x = 0.8$, we find that these lines do form a *tangent* to both curves. The negative exponents behave therefore *just the same* as positive exponents.

f(x) = trigonometric function. For the case where y is the function of an *angle*, we may have a relationship such as the **sine function**,

$$y = \sin\theta$$

(We use using the Greek letter θ here to denote that the independent variable involves an angle; however, one will frequently see in the literature simply $y = \sin x$.)

The sine of an angle is defined as "side opposite over hypothenuse". If we set the hypothenuse equal to 1, then the sine of θ is just equal to the side opposite. Moreover, if we consider the unit circle to overlie a grid of x-y coordinates, then the side opposite becomes equal to y. This is shown in Fig. 9-3.

If we increase θ by $d\theta$, then we have that

$$y + dy = \sin(\theta + d\theta)$$

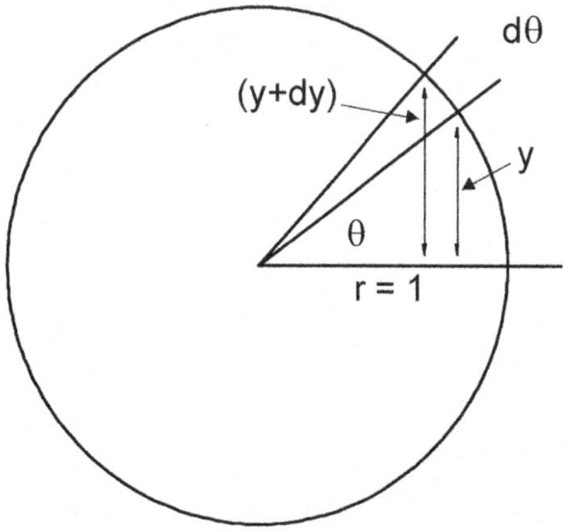

Fig. 9-3 Relationship between dy and d θ

If the preceding eq[illegible] θ, i[illegible] [illegible] equation, we find that,

$$dy = [illegible] \, d\theta)$$

Moreover, trigono[illegible] that [illegible] s subtracted from the sine of another, the resulta[illegible] cal[illegible] the [illegible]

$$[illegible] - \sin [illegible] \frac{[illegible]}{[illegible]} [illegible] \frac{M - N}{2}$$

Setting $M = \theta - d\theta$ [illegible] in this [illegible]

$$[illegible] = 2 \cos \frac{[illegible] + \theta}{2} [illegible] \frac{[illegible] - \theta}{2}$$

$$dy = [illegible] + \frac{1}{2} [illegible] \frac{[illegible]}{2} \, d\theta$$

Now as $d\theta \to 0$, the [illegible] becomes [illegible]ibly [illegible] compared to θ itself. But for the value of the *sine* of [illegible] know fr[illegible] [illegible]metr[illegible] becomes very small, the value of $\sin x$ becomes ve[illegible] [illegible] ng j[illegible] x, so [illegible]

$$[illegible]) = [illegible]$$

We have then that

$$[illegible] \cos [illegible]$$

$$[illegible] \cos \theta$$

$$so [illegible] \cos \theta, [illegible] = \sin [illegible] \quad \frac{[illegible]\sin \theta)}{d\theta} = \cos \theta$$

In English, we wou[illegible] first [illegible] of [illegible] th respect to θ is $\cos \theta$. We saw wit[illegible] abo[illegible] first [illegible] of the function, at a certain x, was in fact the slop[illegible] nt li[illegible] ve [illegible] e. Is this true with trigonometric func[illegible] rivat[illegible] is [illegible] values of $\cos \theta$ for the different degrees of [illegible]

Degs θ	cos θ
30	0.866
45	0.707
60	0.500
75	0.258

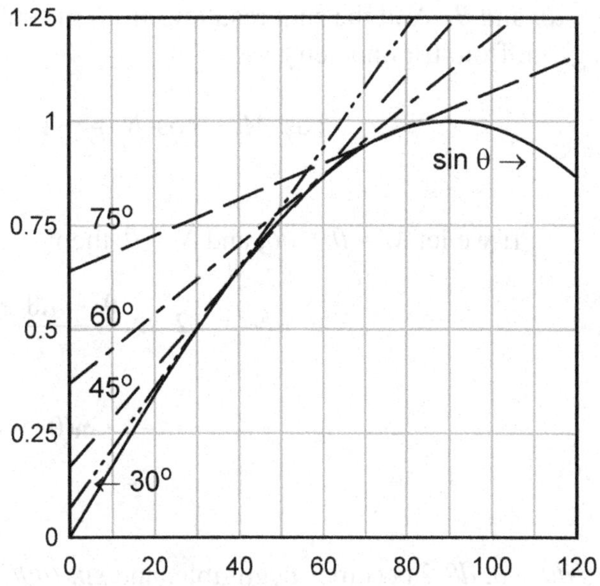

Fig. 9-4 Tangent slope of sine curve vs. cos θ

In Fig. 9-4, we have drawn lines with thesc slopes, and we see that they are indeed tangent to the sine curve at the appropriate points. That is, the line of slope *0.866* is tangent to the sine curve at *30º*, the line of slope *0.707* is tangent at *45º*, and so on. Therefore, the idea that the first derivative of a function represents the slope of the tangent line at the particular *x* chosen holds true also for trigonometric functions.

Let us look next at the differentiation of the **cosine function**. Cosine *θ* is defined as "side adjacent/ hypotenuse", and if the hypotenuse is equal to one (as in Fig. 9-3), then

$$x = \cos\theta$$

However, we wish to keep *y* as the dependent variable, and therefore we invoke the "cosine curve" (as seen in Fig. 9-5) where

$$y = \cos\theta$$

And if we increase *θ* by *dθ*, we will have

$$y + dy = \cos(\theta + d\theta)$$

Then subtracting the first equation from the second shows that

$$dy = \cos(\theta + d\theta) - \cos\theta$$

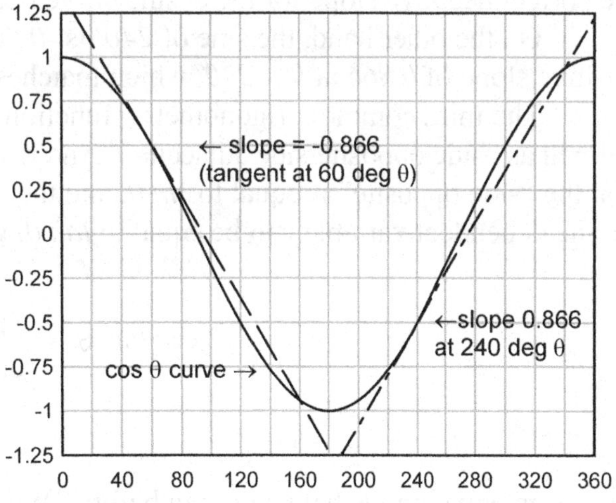

Fig. 9-5 Tangent of -sin derivative to cosine curve

Thus the differential change in *y* is equal to the difference between the cosines of the two angles,

$\theta + d\theta$ and θ. And the sum resulting from the subtraction of the cosines of two different angles is given from trigonometry as

$$\cos M - \cos N = -2 \sin \frac{M + N}{2} \cdot \sin \frac{M - N}{2}$$

If we let $M = \theta + d\theta$, and $N = \theta$, then

$$dy = -2 \sin \frac{\theta + d\theta + \theta}{2} \cdot \sin \frac{\theta + d\theta - \theta}{2}$$

$$= -2 \sin(\theta + \frac{d\theta}{2}) \cdot \sin(\frac{d\theta}{2})$$

As $d\theta \rightarrow 0$, $d\theta/2$ becomes negligible, and $\sin(d\theta/2)$ goes to $d\theta/2$, as was the case in the sine derivative above. This gives us

$$dy = -2 \sin(\theta) \cdot \frac{d\theta}{2}, \textit{ or } \frac{dy}{d\theta} = -\sin \theta, \textit{ and since } y = \cos \theta, \ \mathbf{\frac{d(\cos \theta)}{d\theta} = -\sin \theta}$$

Therefore, the derivative of $\cos \theta$ with respect to θ is equal to $-\sin \theta$.

Notice the fact that the derivative is *negative* "keeps the trigonometry straight." The slope of the cos θ curve from *0* to *180* degrees is negative. But the value of the sine of 60° (*0.866*) is a positive number. However, since the derivative is the negative of $\sin 60^o$, this gives the correct negative slope for the cosine curve between *0* and *180 degrees* in Fig. 9-5.

On the other hand, the sine of 240° is *-0.866*, and the derivative of $-\sin \theta$ yields a positive slope of *0.866* at $\theta = 240^o$ which matches Fig. 9-5.

The third common trigonometric function is the **tangent function**. The tangent is defined as "side opposite/side adjacent." And if we look at the unity circle in Fig. 9-3, we see that the "side opposite" is equal to $\sin \theta$, and the "side adjacent" is equal to $\cos \theta$. Thus if we set the dependent variable y to be equal to $\tan \theta$, we have

$$y = \tan \theta = \frac{\sin \theta}{\cos \theta} = \left(\frac{u}{v}\right)$$

This is an equation we have *not seen* before. We have one function, $\tan \theta$, being equal to a second function, $\sin \theta$, which is divided be a third function, $\cos \theta$. How we add and subtract, or multiply and divide functions is treated in a later section of this Chapter. We will move ahead here, however, and apply the appropriate expression for the division of two functions, which is,

$$\frac{dy}{d\theta} = \frac{v\frac{du}{d\theta} - u\frac{dv}{d\theta}}{v^2}$$

In the present case, the first variable, v, is equal to $\cos\theta$, and its derivative, $dv/d\theta$, is $-\sin\theta$. The second variable, u, is equal to $\sin\theta$, and its derivative, $du/d\theta$, is $\cos\theta$. If we substitute these values into the original expression, we have

$$\frac{dy}{d\theta} = \frac{\cos\theta \cdot \cos\theta - (\sin\theta \cdot -\sin\theta)}{\cos^2\theta}$$

$$= \frac{\cos^2\theta + \sin^2\theta}{\cos^2\theta} \quad \textit{and } \cos^2\theta + \sin^2\theta = 1$$

$$\textit{so, } \frac{dy}{d\theta} = \frac{d\tan\theta}{d\theta} = \frac{1}{\cos^2\theta} = \sec^2\theta, \quad \textit{where } \sec\theta = \textit{secant } \theta$$

And, since $y = \tan\theta$, the derivative of $\tan\theta$ with respect to θ is $1/\cos^2\theta$ or $\sec^2\theta$. The slopes corresponding to $1/\cos^2\theta$ are plotted for $\theta = 60^o$, 75° and 81° in Fig. 9-6. And we see that the derivative lines are tangent to the tangent curve at these points. This is once again the characteristic behavior of the first derivative.

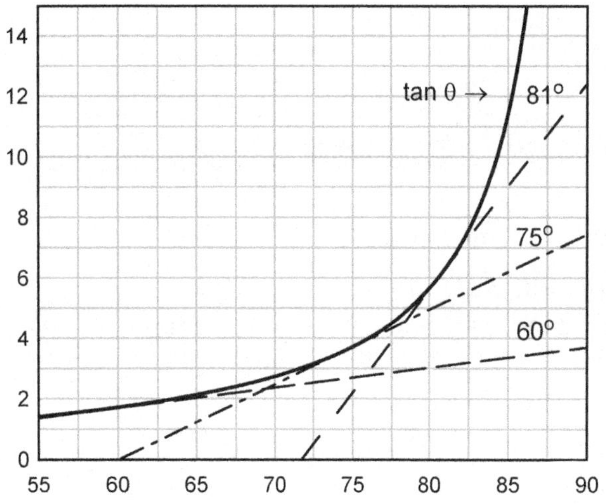

Fig. 9-6 First derivative slope to tangent curve

These last sections illustrate the differentiation of the common trigonometric functions of sine, cosine and tangent. Math handbooks have tables of the derivatives of other trigonometric functions including inverse functions, arc functions, and hyperbolic functions. A discussion of these derivatives is beyond the scope and intent of this tutorial.

However, we will look at the differentiation of the arctangent function, since the derivative of this function was introduced in the section on power series calculations (pg. 136). We noted there that the derivative of the arctangent function was [illegible] $= \frac{1}{1 + x^2} \, dx$

And from power se[illegible] [illegible]ions, [illegible] [illegible] that

$\frac{1}{1 + x^2} =$ [illegible] . . .

We could thus appr[illegible] value [illegible] arctange[illegible] Fig. 8-1) for values of x within the *limits* of $x = -1$ [illegible] the int[illegible] the [illegible]series,

arcta[illegible] x^2 [illegible]

[illegible] $\frac{x^3}{3} +$ [illegible] (*integration constant* $= 0$)

But how we knew [illegible] x^2) was [illegible] [illegible]tive [illegible] [illegible]ngent function was not explained.

This identit[illegible] calc[illegible] is j[illegible] of the differentiation of the arctangent functio[illegible] [illegible] [illegible]ion [illegible] [illegible]gent [illegible]

arct[illegible] [illegible] $^{-1}$ y)

an[illegible] *an*[illegible] [illegible] *then* $y = \tan \theta$

From our work abo[illegible] that [illegible] inte[illegible] [illegible]s in determining the first derivative of $y =$ *ta*[illegible] [illegible] identi[illegible]

[illegible] $\frac{[illegible]^2\theta + [illegible]}{\cos[illegible]}$

This is the derivati[illegible] [illegible]umer[illegible] of *ta*[illegible] [illegible]spect to the θ (the angle itself). It represents the *ch*[illegible] [illegible] result[illegible] inf[illegible] [illegible]small change in θ. On the other hand, when we stu[illegible] [illegible]tive [illegible] [illegible]gent [illegible] we wish to find the change in θ that is produced by [illegible] [illegible] chan[illegible] [illegible] do [illegible] [illegible]vert" the identity above, and simplify it by dividi[illegible] [illegible]rator [illegible] [illegible]inat[illegible] [illegible]ht side by $cos^2\theta$, which gives

$$\frac{d\theta}{dy} = \frac{\cos^2\theta}{\cos^2\theta + \sin^2\theta}$$

$$= \frac{\dfrac{\cos^2\theta}{\cos^2\theta}}{\dfrac{\cos^2\theta}{\cos^2\theta} + \dfrac{\sin^2\theta}{\cos^2\theta}} = \frac{1}{1 + \tan^2\theta}$$

And then we note that since

$y = \tan\theta$, *we can square both sides and show that,* $y^2 = \tan^2\theta$

and so $$\frac{d\theta}{dy} = \frac{1}{1 + y^2}$$

The derivative of θ with respect to y is therefore $1/(1 + y^2)$. (Some may be puzzled because the original differentiation was given in terms of x. But it does not matter if the independent variable is called x or y; the form of the differential identity stays the same.)

Fig. 9-7 shows a graph of the arctangent function. For an angle of *50°*, for example, the tangent is read from the curve as *1.2*. (The calculator value for tan *50°* is *1.192*.)

If we consider an angle of *75°*, the tangent value from the curve is *3.75*. And therefore the value of the derivative at *75°* is given by

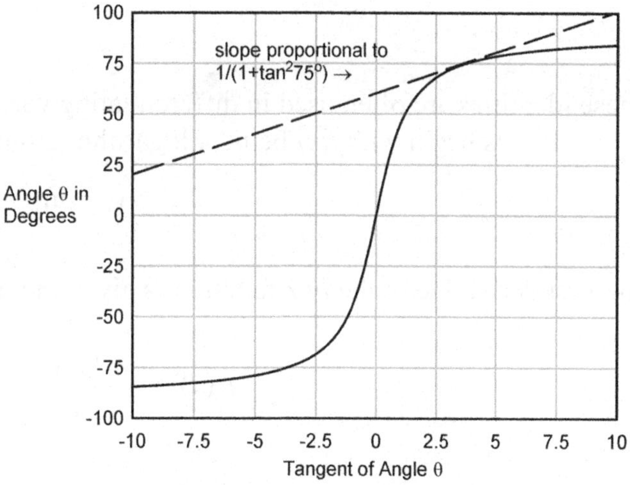

Fig. 9-7 Derivative of arctangent as slope to curve at 75 deg

$$d\theta = \frac{1}{1 + \tan^2 75^o}\, dy$$

$$= \frac{1}{1 + 3.75^2}\, dy$$

$$d\theta = 0.0664\, dy$$

This is the *instantaneous* rate of change at *75°*, and it represents the change in tangent θ that would be produced by an infinitesimally small change in the angle θ .

And, if we draw a straight line whose slope is proportional to the derivative at 75°, we find that this line is tangent to the arctangent curve at, and only at, *75°*. So we see once again the characteristic relationship between the first derivative of a function, and the tangent to the curve generated by that function.

F(x) = ln(x) or e ^{x}. Logarithms and exponents are related functions. For example, in the expression

$$y = \log_{10} x$$

if x = *100*, then y = *2* because $10^2 = 100$. But we could also write the expression in the exponential form, i.e.,

$$x = 10^2, \quad \textit{or in general form} \quad x = 10^y$$

This expression can be written, in addition, as

$$x = 10^{\log_{10} x}$$

And for the natural logarithm, we have similar expressions,

$$y = \ln_e x \qquad x = e^y \qquad x = e^{\ln_e x}$$

These identities are often used in differentiating various logarithmic and exponential functions.
Let us begin with *f(x)* being a logarithmic function. The simplest such function is

$$y = \ln x$$

The first derivative of the *ln x* function is given in handbooks as

$$f'(x) = \frac{d(\ln x)}{dx} = \frac{dy}{dx} = \frac{1}{x}$$

This derivative is found as follows: The logarithmic function can also be written as

$$e^y = x$$

And if we take the derivative of *both* sides with respect to x, the result is

$$e^y \frac{dy}{dx} = \frac{d(x)}{dx} = 1$$

There are two important points here. We want to know how the *left* side of the equation changes with respect to x, but the exponential variable is y. Therefore we need to know how y changes with x, so we introduce the differential dy/dx on the left side. Secondly, the derivative of x with respect to x is just equal to *1*, which becomes the right side of the equation.

As we look at this last equation, we recall from the preceding equation that $e^y = x$, so we make the substitution and find

$$x \frac{dy}{dx} = 1 \quad or \quad \frac{dy}{dx} = \frac{1}{x}, \qquad and\ since\ y = \ln x, \quad \frac{\mathbf{d(ln\ x)}}{\mathbf{dx}} = \frac{\mathbf{1}}{\mathbf{x}}$$

Therefore the first derivative of the function, *ln(x)*, is equal to $1/x$, a result that one might not guess *a priori*.

The second derivative of *ln x* can be found by the differentiation procedure for power terms, i.e.

$$\frac{d(x^n)}{dx} = \frac{1}{n} x^{n-1}$$

And since $1/x = x^{-1}$, we have

$$f^{\prime\prime}(\ln x) = \frac{d(\frac{1}{x})}{dx} = -\frac{1}{x^2}$$

The third, fourth, etc. derivatives of *ln x* can be determined simply by an extension of the differentiation procedure for power terms.

The logarithmic function may be not just of x, but of an **expression** involving x such as

$$f(x) = y = \ln(4x + 3)$$

We can differentiate this function by the **chain rule** (see following section). Setting the *inner function* to be $(4x + 3)$, we have [illegible] $x + 3$ [illegible] $\frac{d[illegible]}{dx}$ [illegible] $3) = 4$

And then setting the [illegible] *ion* eq[illegible] we [illegible]

[illegible] $= \ln$ [illegible] $\frac{1}{4x + 3}$

and then ch[illegible] $\times \frac{d[illegible]}{dx}$ [illegible] $\frac{d(\ln(4x + 3))}{dx} = \frac{4}{4x + 3}$

We can also [illegible] logarith[illegible] [illegible]ber [illegible] [illegible]nt **base**. For example, the function may be

[illegible] $\log_{(10)}$ [illegible]

And suppose we w[illegible] the deri[illegible] this l[illegible] We know the derivative of the *natural* logarithm [illegible] $= ln$ [illegible] is *no*[illegible] [illegible]tive to a different base, which is base *10* in this case.

However, th[illegible] between [illegible]garithm [illegible]ral base, *e*, and a logarithm to any other base is sim[illegible] *[illegible]nt*, e.g.

[illegible] *of* [illegible] $= 1$

[illegible] $\ln_{(e)}$ [illegible] $\times$ [illegible] *any number*

and [illegible] $\ln 10$ [illegible]

If we generalize the [illegible] this eq[illegible] [illegible]n we [illegible]

[illegible] $\ln_{(10)} =$ [illegible]

Returning to our ori[illegible]n, we [illegible]

$$f(x) = \log_{(10)} x = \frac{\ln x}{\ln 10} = \frac{1}{2.303} \times \ln x$$

$$\textit{and } f'(x) = \frac{1}{2.303} \frac{d(\ln x)}{dx} = \frac{1}{2.303} \frac{1}{x}, \quad or \quad \frac{dy}{dx} = \frac{1}{\ln a} \frac{1}{x} \;\; \textbf{(for base a)}$$

In the realm of **exponentials**, one of the most interesting derivatives is that of the function,

$$y = e^x, \text{ where } y' = e^x$$

That is, the derivative of the function, and in fact *all* derivatives of the function, is just the function *itself.* That seems strange to students, but it can be explained from the derivation. Recall from the original definition of *e* by Euler (p.38) that

$$e = \lim_{x \to \infty} (1 + \frac{1}{x})^x, \text{ and if we set } x = \frac{1}{\Delta x} \text{ where } \Delta x \text{ is small,}$$

$$e \cong (1 + \Delta x)^{\frac{1}{\Delta x}}, \text{ and taking both sides to the } \Delta x \text{ power, } \mathbf{e^{\Delta x} = (1 + \Delta x)}$$

We can now use the "Δy/Δx" method of differentiation. If we consider $y = e^x$ to be plotted as an *x-y* graph, and we change *x* by Δx (i.e., $x_1 - x_0 = \Delta x$), then $\Delta y = e^{x + \Delta x} - e^x$, and our differentiation from the $\Delta y/\Delta x$ ratio is given by

$$f'(x) = \lim_{\Delta x \to 0} \frac{e^{x + \Delta x} - e^x}{\Delta x} = \lim_{\Delta x \to 0} \frac{e^x e^{\Delta x} - e^x}{\Delta x}$$

$$= \lim_{\Delta x \to 0} \frac{e^x(e^{\Delta x} - 1)}{\Delta x} = \lim_{\Delta x \to 0} \frac{e^x(1 + \Delta x - 1)}{\Delta x} \;\; (\text{since } e^{\Delta x} = 1 + \Delta x)$$

$$= \lim_{\Delta x \to 0} \frac{e^x \Delta x}{\Delta x}, \quad so \;\; \mathbf{f'(x) = \lim_{\Delta x \to 0} e^x = e^x}$$

A second, and perhaps easier derivation, can be had by recalling that e^x can be defined by the Taylor series expansion,

$$e^x = \sum_{n=0}^{\infty} \frac{x^n}{n!} = 1 + x + \frac{x^2}{2!} + \frac{x^3}{3!} + \frac{x^4}{4!} + \cdots$$

We have then that the first derivative of e^x is equal to

$$\frac{d(e^x)}{dx} = 0 + \frac{1}{1} + \frac{2}{1\cdot 2}x + \frac{3}{1\cdot 2\cdot 3}x^2 + \frac{4}{1\cdot 2\cdot 3\cdot 4}x^3 + \frac{5}{1\cdot 2\cdot 3\cdot 4\cdot 5}x^5 + \cdots$$

$$= (0) + 1 + x + \frac{x^2}{2!} + \frac{x^3}{3!} + \frac{x^4}{4!} + \cdots$$

That is, the series that we obtain by differentiating the *original* series for e^x is just the same as the original series, but "moved over" one term to the left. And, since the Taylor series consists of an *infinite* number of terms, each new differentiation will just yield a new series that is equal to e^x. So $d(e^x)/dx$ will always equal e^x.

The numerical value of e^x may seem nebulous, but it is just $(2.718..)^x$, where x can take on various values. In Fig. 9-8, we see that, when $x = 1$, the derivative is equal to $2.718...$, and that a line of this slope is tangent to the e^x curve at the point, $x = 1$. Similarly, the value of the derivative at $e^{2.5}$ is equal to $12.182...$, and a line of this slope is tangent to the e^x curve at $x = 2.5$. Once again, we find that the first derivative is tangent to the curve generated by the function.

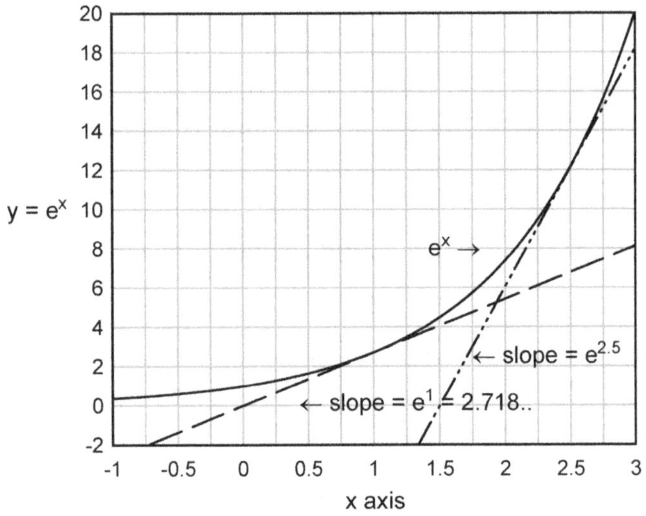

Fig. 9-8 Tangent slope to curve is equal to e to the x power

We have seen here how the derivatives are calculated for a few different classes of functions. In a regular calculus class, students are introduced to more types of functions, and are taught a greater array of methods (or "tricks") for differentiating functions. The general approach is the same as demonstrated here, however. Students also have the advantage nowadays that they can search for the derivatives of given functions on the Internet. There are several Internet sites such as *Calc101* that will differentiate virtually any function, and provide a description of how the derivative was obtained.

Arithmetic operations with functions

A function may be produced by the addition or subtraction, or multiplication or division, of two other functions. The derivative of such functions is given in tables of derivatives as

function	*derivative*
$y = u + v$	$dy = d(u + v) = du + dv$
$y = uv$	$dy = d(uv) = udv + vdu$
$y = \frac{u}{v}$	$dy = d(\frac{u}{v}) = \frac{vdu - udv}{v^2}$

In these tables, y, u and v are all functions (but *different* functions) of x, and although it is not done, we could write them as $y(x)$, $u(x)$ and $v(x)$ for maximum clarity if we wish.

Addition and subtraction. These operations can be defined from "first principles." In addition or subtraction, if u and v are changed by du and dv, then y will change by dy. This is shown by

$$y + dy = u + du + v + dv = (u + v = y) + du + dv, \quad or \ \mathbf{dy = du + dv}$$

And since these variables are all functions of x, we can write

$$\frac{dy}{dx} = \frac{du}{dx} + \frac{dv}{dx}$$

to indicate that they each vary with x. For example, let $u = x^2 + a$, and $v = 2x^3 + b$, then

$$\frac{dy}{dx} = \frac{d(x^2 + a)}{dx} + \frac{d(2x^3 + b)}{dx}$$

$$= 2x + 3 \cdot 2x^2$$

$$or \quad dy = (6x^2 + 2x)\ dx$$

Moreover, if we look at the original equation, we see that y can be the sum of any number of functions including those which are negative, and the derivative will be

w [illegible] dv [illegible] $- dw + dz$

Multiplicati[illegible] **operatio**[illegible] **lied a**[illegible]

[illegible] $\times$ [illegible]

and, for the change [illegible] dy [illegible]

dy = [illegible] $\times$ [illegible]

= [illegible] $(du \cdot dv = 0)$

dy = [illegible]

Since du and dv are [illegible] esimal [illegible] in old [illegible] atics they were called "infinitesimals"), th[illegible] *esp*[illegible] nitesi[illegible] n be neglected. An infinitesimal multi[illegible] n-in[illegible] car[illegible] idered as equal to zero, however, because [illegible] esim[illegible] as [illegible] nity.

If we refer [illegible] 2-1, w[illegible] es th[illegible] iation of $y = x^2$, we see that the secant to the curve [illegible] ome [illegible] d Δx, and hence Δy, becomes very small. It is on[illegible] and Δ[illegible] qual [illegible], that we can say

[illegible] *whi*[illegible] *deri*[illegible] $= f(x^2)$,

and [illegible] *of* [illegible] ***t at x***

Note that although [illegible] the [illegible] ro, [illegible] dx can be zero, for then we would not have the [illegible] $+ dy)$ [illegible] t [illegible] e slope of the tangent.

This has lea[illegible] ment b[illegible] atici[illegible] years as to whether dx, dy and so on should be con[illegible] ry sm[illegible] rs, or [illegible] that approach, but never reach, a limit of 0. In this t[illegible] ill mo[illegible] dx, dy, [illegible] ing infinitely small.

We will de[illegible] deriv[illegible] for m[illegible] by an example. Let $u = x^2 + a$, and $v = 2x^3 + b$, [illegible] then b[illegible] ultipl[illegible] at

y = [illegible] $+ a)$ [illegible] $=$ [illegible] $+ bx^2 + ab$

so [illegible] $+$ [illegible]

Using the derivative rule for multiplication gives

$$\frac{dy}{dx} = u\frac{dv}{dx} + v\frac{du}{dx} = (x^2 + a)(6x^2) + (2x^3 + b)(2x)$$

$$= 6x^4 + 6ax^2 + 4x^4 + 2bx$$

$$\frac{dy}{dx} = 10x^4 + 6ax^2 + 2bx$$

So we obtain the same answer either by multiplication and then differentiation, or by "differentiation and then multiplication."

Division. In developing the derivative rule (or formula) for division, we make use of the multiplication rule. If we have

$$y = \frac{u}{v} \quad \textit{this means that} \quad u = y \times v$$

From the multiplication rule, we know that

$$du = vdy + ydv,$$

$$\textit{and solving for dy gives,} \quad dy = \frac{du - ydv}{v}$$

and then substituting the original definition for y $(= u/v)$, we have

$$dy = \frac{du - \frac{u}{v}dv}{v} = \frac{duv - udv}{v^2}, \quad \textit{and so} \quad \mathbf{dy} = \frac{\mathbf{vdu - udv}}{\mathbf{v^2}}$$

The use of the division rule was demonstrated earlier (pg. 161) in determining the derivative of the tangent function, $y = \tan\theta = \sin\theta / \cos\theta$.

The chain rule

This is one of the most useful rules (or "tricks") in calculus. It allows us to differentiate functions that would otherwise be difficult. Essentially it involves splitting the function under study into two functions, one being the "inside function" and the other the "outside function". Consider the example,

$$y = (x^3 + a^{-\frac{1}{2}})^{\frac{5}{2}}$$

We name the inside function *u*, and define it as being

$$u = (x^3 + a^{-\frac{1}{2}})$$

Following this, we define the so-called "outer function" as being

$$y = (u)^{\frac{5}{2}}, \quad \textit{with then} \quad \frac{dy}{du} = \frac{5}{2}(u)^{\frac{3}{2}}$$

We now differentiate the "inner function" *u* with respect to *x*, i.e.,

$$u = (x^3 + a^{-\frac{1}{2}}) \quad \textit{and} \quad \frac{du}{dx} = 3x^2$$

And then we "chain" these two derivatives together to get

$$\frac{dy}{dx} = \frac{dy}{du}\frac{du}{dx} = \frac{5}{2}(u)^{\frac{3}{2}} \times 3x^2$$

and substituting the original value for *u* back into the equation gives

$$\frac{dy}{dx} = \frac{dy}{(du)}\frac{(du)}{dx} = \frac{5}{2}(x^3 + a^{-\frac{1}{2}})^{\frac{3}{2}} \times 3x^2, \quad so \quad \frac{dy}{dx} = \frac{15}{2}x^2(x^3 + a^{-\frac{1}{2}})^{\frac{3}{2}}$$

Notice in the last equation that the *du's* "cancel out", and we are left just with the derivative of the original function *y* with respect to *x*, or

$$y = (x^3 + a^{-\frac{1}{2}})^{\frac{5}{2}} \quad and \quad \frac{dy}{dx} = \frac{15}{2}x^2(x^2 + a^{-\frac{1}{2}})^{\frac{3}{2}}$$

The chain rule can be applied to problems that involve more than two differentials. Here is an example taken from the literature (pg.99, *Calculus Made Easy (1998 ed.)*, S. P. Thompson and M. Gardner, St. Martin's Press, New York, 1998) which shows how this is done. We seek to find the first derivative of the function,

$$y = \frac{x}{b}\sqrt{(a - x)x}$$

$$let\ u = \sqrt{(a - x)x}\ ,\quad then\ y = \frac{x}{b}u$$

$$\frac{\mathbf{dy}}{\mathbf{dx}} = \frac{x}{b}\frac{du}{dx} + \frac{u}{b}\ (multiplication\ rule),\quad set\ w = (a - x)x,\quad then\ u = w^{\frac{1}{2}}$$

$$\frac{\mathbf{du}}{\mathbf{dw}} = \frac{1}{2}w^{-\frac{1}{2}} = \frac{1}{2w^{\frac{1}{2}}} = \frac{1}{2u}$$

$$w = (a - x)x = ax - x^2\ ,\quad so\ \frac{\mathbf{dw}}{\mathbf{dx}} = a - 2x$$

Now we can consolidate the equations for *dy/dx*, *du/dw* and *dw/dx*, i.e.,

$$\frac{du}{dw}\times\frac{dw}{dx} = \frac{du}{dx} = \frac{1}{2u}\times(a - 2x) = \frac{a - 2x}{2u}$$

$$and\ \frac{dy}{dx} = \frac{x}{b}\frac{du}{dx} + \frac{u}{b} = \frac{x(a - 2x)}{2bu} + \frac{u}{b}$$

$$since\ u = \sqrt{(a - x)x}\ ,\ then\ \frac{dy}{dx} = \frac{x(a - 2x)}{2b\sqrt{(a - x)x}} + \frac{\sqrt{(a - x)x}}{b} = \frac{x(3a - 4x)}{2b\sqrt{(a - x)x}}$$

After considerable manipulation of the chain rule, and using three differential expressions, we ultimately find that for

$f(x) =$ [illegible] $x)x$, [illegible] *deri*[illegible] $(x) = \dfrac{x(3a - 4x)}{2b\sqrt{(a - x)x}}$

Partial di[illegible]n

A depende[illegible] be th[illegible] of [illegible]ne independent variable. We indicate this condit[illegible] [illegible]ality

[illegible] $(u,$ [illegible]

[illegible] mor[illegible] *etc.*)

To find the change [illegible]ts fro[illegible] in [illegible] uses **partial differentiation** where one indepen[illegible] is trea[illegible] function is differentiated with respect to the other [illegible] pro[illegible] re[illegible]erse order. For example, assume that

[illegible]

[illegible]

[illegible]

The subscripts *v* an[illegible] that [illegible] trea[illegible]ants in the indicated equations. The term "partial d[illegible]" aris[illegible] the [illegible]ation is differentiated only *partially* (i.e., with [illegible] y or[illegible] each of the differential equations.

To indicate [illegible]entiati[illegible] mat[illegible]e *d* by the symbol, ∂, e.g.,

[illegible] *mes* [illegible] *nd* [illegible] *omes* $\dfrac{\partial y}{\partial v} = u$

And if we substitut[illegible]ies [illegible] orig[illegible]ential equations, we find that

$dy_v =$ [illegible] *nd* [illegible]

These equations represent the *values* of the changes in y that occur because of the individual changes in v and u. Summing up the values of these two changes, we find the total change in y (or its *total differential*) to be given by

$$dy = \frac{\partial y}{\partial u}du + \frac{\partial y}{\partial v}dv$$

The total differential is also sometimes written as

$$dy = \left(\frac{dy}{du}\right)du + \left(\frac{dy}{dv}\right)dv \quad or \quad dy = \left(\frac{dy}{du}\right)_v du + \left(\frac{dy}{dv}\right)_u dv$$

One of the easier ways to visualize partial derivatives is to consider the three-dimensional figure that results from an arbitrary function such as,

$$z = 2x^2 - 3xy + y^2$$

For the coordinates of, say, $x = 2, y = 1$, we find that $z = 3$. The value, or "height", of z above the point, *2,1*, on the two-dimensional *x-y* plane is thus *3*. If we repeat this calculation for all possible values of x and y, the resultant values of z would form a "roof" over some type of three-dimensional form.

Now let us follow the rules for partial differentiation and differentiate the function first with respect to x holding y constant, and then with respect to y holding x constant. This gives

$$\frac{\partial z}{\partial x} = 4x - 3y$$

$$and \quad \frac{\partial z}{\partial y} = -3x + 2y$$

Evaluating these derivatives at the point, *2,1,* we then have

$$\frac{\partial z}{\partial x} = 4(2) - 3(1) = 5 \quad (tangent\ slope\ parallel\ to\ x\text{-}axis)$$

$$\frac{\partial z}{\partial y} = -3(2) + 2(1) = -4 \quad (tangent\ slope\ parallel\ to\ y\text{-}axis)$$

The resultant total differential change in z is thus

$$dz = 5dx - 4dy$$

If we assume a small change in x (say, $dx = 0.005$) and in y ($dy = 0.003$), the approximate change in z would be

$$dz = 5(0.005) - 4(0.003) \approx 0.013 \quad (\textit{approximate change in } z)$$

And, as the *dx* and *dy* changes become infinitesimally small, *dz* becomes the total differential change in *z*, or

$$dz = \frac{\partial z}{\partial x}dx + \frac{\partial z}{\partial y}dy$$

To extend this idea, one may think of the plane, $y = 1$, (i.e., where y is constant) as intersecting the dome of the three-dimensional figure. This intersection generates a two-dimensional *curve* which progresses parallel to the x direction, and the tangent *line* to this curve is given by

$$\frac{\partial z}{\partial x} = 4x - 3y$$

which is the *partial* derivative of $z(x,y)$ when y is a constant. A similar treatment when $x = 2$ gives

$$\frac{\partial z}{\partial y} = -3x + 2y$$

which is the partial derivative of $z(x,y)$ for a constant value of x.

These two partial derivative tangent lines lie in the **tangent plane** which just *touches* our three-dimensional figure at the point in space of $x,y,z = 2,1,3$. The plane tangent contains the original point in space $(2,1,3)$ as well as *all* points that occur on the two partial derivative tangent lines. We can therefore specify a point on *each* partial tangent line, and combine these points with the initial point, to calculate the equation of a "plane (i.e., the tangent plane in this case) through three points in space."

Consider three points in space. If they are not in linear arrangement, they define a single unique plane in space. One draws the axis of a plane through two points, and then rotates the plane until it contacts the third point. This gives three points on the same plane. We can then use the coordinates of these three points to calculate the equation for the plane, which is a linear equation of the form,

$$ax + by + cz = d$$

The original point has the coordinates of $\boldsymbol{P_1} = \boldsymbol{2,1,3}$. The slope of the x tangent line at this point is 5 units of z per 1 unit of x, so if x increases by 1, z increases by 5. The coordinates of our second point are therefore *3(2+1),1(y is constant),8(3+5)* or $\boldsymbol{P_2} = \boldsymbol{3,1,8}$. The y tangent line has a slope of -4 z to 1 y, so if y increases by 1, z decreases by -4. This gives the coordinates of the third point as *2(x is constant), 2 (1+1),-1 (3-4),* or $\boldsymbol{P_3} = \boldsymbol{2,2,-1}$.

To find the equation of the tangent plane, we must find the *coefficients* which satisfy the three linear equations,

$$\begin{aligned} &1) \quad 2a + b + 3c = d \\ &2) \quad 3a + b + 8c = d \\ &3) \quad 2a + 2b - c = d \end{aligned}$$

We approach this problem by Gaussian elimination (pg. 64), and subtract eqn 1) from eqn 3) to obtain

$$0 + b - 4c = 0, \quad or \quad \boldsymbol{b = 4c}$$

And if we multiply eqn 2) by 2, and eqn 1) by 3, and then subtract the resultant eqn 2) from the resultant eqn 1), we find

$$0 + b - 7c = d, \quad or \quad b - 7c = d$$

$$and\ then\ because\ b = 4c$$

$$-3c = d, \quad or\ \boldsymbol{c} = \frac{\boldsymbol{-d}}{\boldsymbol{3}}$$

We can next solve for a in terms of d using *eqn 3*,

$$2a + 2b - c = d \quad (eqn\ 3)$$

$$so\ 2a = d - 2b + c = d - 2\left(\frac{-4d}{3}\right) - \frac{d}{3} = \frac{10d}{3}$$

$$and\ 2a = \frac{10d}{3}, \quad giving\ \boldsymbol{a} = \frac{\boldsymbol{5d}}{\boldsymbol{3}}$$

This gives us the co[illegible] he li[illegible] [illegible]w introduce the variables, and we have that

$$\left|\frac{5d}{3}\right. \quad [illegible]$$

and if we let [illegible] **5**[illegible] [illegible] *ation of the tangent plane*)

We can then substit[illegible] ordin[illegible] origi[illegible] verify that this equation does in fact specify the t[illegible]

[illegible]ne [illegible] $5x$ [illegible] 3

[illegible] $= 2,1$ [illegible] $(2$ [illegible] $= 3$

[illegible] $= 3,1$ [illegible] $(3$ [illegible] $= 3$

[illegible] $= 2,2$ [illegible] $= 3$

Approach [illegible] fferenti[illegible]

Partial diffe[illegible] basic[illegible] [illegible]ntiation of functions having a single independent [illegible] ll but [illegible] inde[illegible] iables are treated as *constants*, and the function is [illegible] riated [illegible] variable. The process is repeated for all of [illegible] ent va[illegible] ifferentiating single variable functions, but with [illegible] ants" [illegible] dif[illegible], the total change in the dependent variable [illegible] to th[illegible] differentials,

$$dz = [illegible] + \left(\frac{\partial z}{\partial v}\right)_{x,y,u} dv + \cdots$$

An examp[illegible] se of [illegible] liffe[illegible]

Partial deri[illegible] inte[illegible] the[illegible] calculate the volume of three-dimensional figure[illegible] o-fol[illegible] le i[illegible] calculation of the changes in volume of a cone th[illegible] he co[illegible] changes. The volume of a cone is given by

$$V = \frac{1}{3}\pi r^2 h$$

$$and \quad \left(\frac{\partial V}{\partial r}\right)_h = \frac{2\pi}{3} r h$$

$$\left(\frac{\partial V}{\partial h}\right)_r = \frac{\pi}{3} r^2$$

The volume of a cone depends on its height and radius at the *same* time. How do you calculate the effect of two independent variables at the same time? The answer is that you "separate them out" by means of the total differential, which we *integrate* (see next section) according to,

$$\Delta V = \int_{V_1}^{V_2} dV = \int_{r_1}^{r_2} \left(\frac{\partial V}{\partial r}\right) dr + \int_{h_1}^{h_2} \left(\frac{\partial V}{\partial h}\right) dh$$

$$= \int_{r_1}^{r_2} \left(\frac{2\pi}{3} r h\right) dr + \int_{h_1}^{h_2} \left(\frac{\pi}{3} r^2\right) dh$$

Here ΔV is the macroscopic change in V that results when we increase r_1 to r_2 or h_1 to h_2. Let $r_1 = 2$, and $h_1 = 4$ initially, and let us then increase these parameters to $r_2 = 4$ and $h_2 = 6$.

The integral of a derivative is often called the *antiderative* because integrating the derivative returns the original function (except for a possible constant of integration). For example,

$$V = \frac{1}{3}\pi r^2 h, \ and\ since\ \frac{d(r^2)}{dr} = 2r, \quad \frac{\partial V}{\partial r} = \frac{\pi}{3}(2r)h$$

$$but\ then \int \left(\frac{\partial V}{\partial r} dr\right) \ just\ equals \ \frac{1}{3}\pi r^2 h = V$$

In the equation above, the symbol ∫ indicates that we are *integrating* the partial differential of V with respect to r. The partial differential of V with respect to h also integrates back to the original function. Therefore, we have that

$$\Delta V = \left[\frac{1}{3}\pi r^2 h(=4)\right]_2^4 + \left[\frac{1}{3}\pi r(=4)^2 h\right]_4^6$$

$$= \frac{4}{3}\pi(4^2 - 2^2) + \frac{1}{3}\pi(4^2)(6 - 4)$$

$$= 50.265 + 33.510 = 83.775 \ units^3$$

In this equation, the *brackets* indicate that we are dealing with an integral. The numbers at the right-hand bracket set the *limits* over which we are to evaluate the integral. Subtracting the value of the integral at its lower limit (e.g., $r = 2$, or $h = 4$) from its value at its higher limit ($r = 4$ or $h = 6$) yields the net value of the integral.

In some cases, the integrals are not indicated by brackets, but simply by vertical lines at the right side of the expression. These lines will have upper and lower numbers however, just as here, to signify the limits over which the integral is to be evaluated.

Note that we start with the constant *h* being equal to *4* (its *lowest* value) in the first integral, and the value of the constant *r* being equal to *4* (its highest value) in the second integral. If we reverse the order, and make the constant *h* equal to *6* in the first integral, and the value of the constant *r* equal to *2* in the second integral, we obtain this answer,

$$\Delta V = 75.398 + 8.377 = 83.775 \ units^3$$

The total volume change, ΔV, is *still* the same, but the values of integrals *1* and *2 are* changed.

This is explained in Fig. 9-9. Integrating from $r = 2$ to $r = 4$ with $h = 4$ produces the cross-hatched region in the left triangle. Integrating from $h = 4$ to $h = 6$ with $r = 4$ yields the diagonally lined region in the left triangle. But, if we integrate from $r = 2$ to $r = 4$ with *h* at *6*, we produce the cross-hatched region in the right triangle. A second integration from $h = 4$ to $h = 6$ with *r* at *2* gives the diagonally lined region in the right triangle.

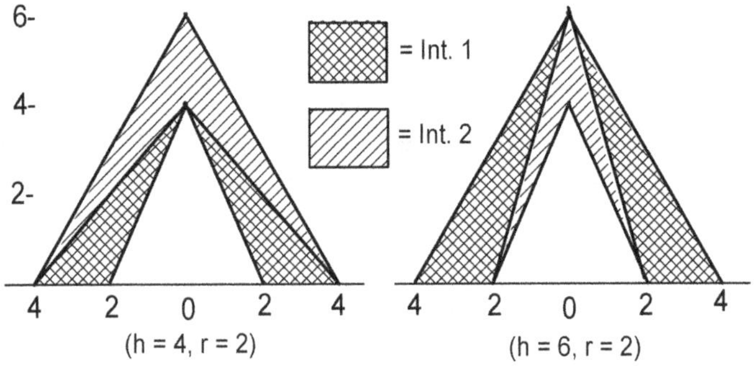

Fig. 9-9 Cross-section of cone showing volumes measured by integrals 1 and 2

The diagonally lined region in the left triangle is clearly greater than that in the right triangle. But the sum of the volumes measured by the two integrals is just equal to ΔV in either case. In each case, we are essentially subtracting the volume of the clear inner cone, V_1 (where $r = 2$ and $h = 4$), from the volume of the outer cone, V_2 (where $r = 4$ and $h = 6$), i.e.,

$$\Delta V = V_2 - V_1 = \frac{1}{3}\pi(r = 4)^2(h = 6) - \frac{1}{3}\pi(r = 2)^2(h = 4)$$

$$= 100.530 - 16.755 = 83.775 \ units^3$$

We will return to the evaluation of partial differentials in the section below on **double integrals**.

Integration

When students look at the Tables of Integrals in Mathematical Handbooks, they see a listing of over *700* different integrals. And they are advised in the introduction "... that it is (still) a fairly uncommon occurrence to find in the table the exact integral desired." This is disheartening.

Integration at the highest level is difficult. The Wolfram Mathematica Online Integrator site (http://integrals.wolfram.com/about/history/) gives a brief history of integration which lists many renowned mathematicians who have studied integration. The first table of integrals was compiled by **C. F. Gauss** (*1777-1855*). And since that time much work has been conducted, even into the 20th century, to extent the scope of the integral tables.

The practical problem of integration has been largely solved, however, by the recent development of computer programs that can automatically differentiate or integrate virtually any function. Several of these programs are available on the Internet. Two well-known sites are *Calc101*, which provides free computer differentiation of functions (and cites for the student the rules involved), and the *Wolfram Integrator*, which lets one do free computer integrations of functions. These web sites let student experiment with differentiation/integration, and also check their home work answers.

An idea of the complexity of such programs is given in the Frequently Asked Questions (FAQ) section of *Calc101*, which notes that their differentiation program uses *15* rules, but that *more* than *300* rules are involved in their integration program.

In this introductory tutorial, we will look only at the basic concept of integration and a few examples of how to integrate simple functions.

The basic concept of integration

The basic concepts of differentiation and integration were introduced on pgs. 25-29, and the roles of the function *f(x)*, the derivative *f'(x)* and the integral *F(x)* are illustrated in Fig. 9-10.

In the prece ... s, we ... hat the function genera ... the ... that the derivative give ... o th ... very value of x. We wil ... how ... measures the area ... e be ... and $x(1)$.

Consider th ... n Fig. ... rea i equal to $1 \times 2 = 2$ ... s area ... al to the *sum* of the 6 in ... o-are ... Conceivably we co ... x a ... million parts. And ... he $1 x$ ... would then be equal ... of th ... million *micro-rect* ... re pr ...

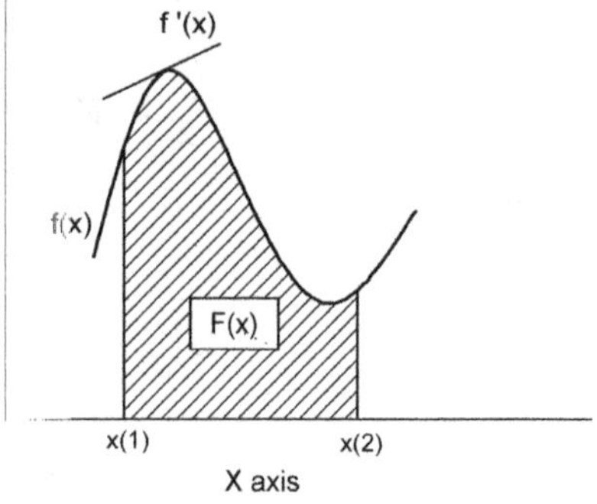

Relation of function, derivative and integral

This calculat ... beca ... is a *rectangle*, and their areas are all equal. How ... that ... he m ... is *curved* rather than flat. How then can we calcul ... the m ... region? The secret ... t tha ... height, y, changes ... efore, function of x, i.e, y ... it is function of x no ma ... valu ... Y a ... or how small x mig ...

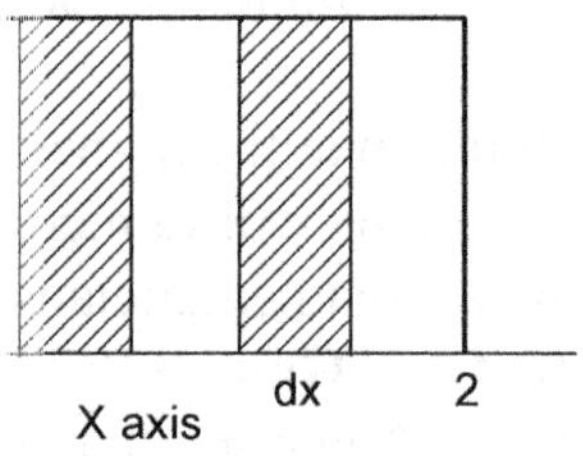

tangle by summation of micro-areas

As x becom ... (or ... magnification bec ... e), a ... curve can be appro ... strai ... line, and the micro- ... hen h ... g. 9- ... sloping straight lin ... g. 9-1 ... If the height y lies ... f dx ... the hatched areas a ... left ... and ... the micro-figure is given by

$= y$ ...

$= f($ ...

then ... $= f($ ...

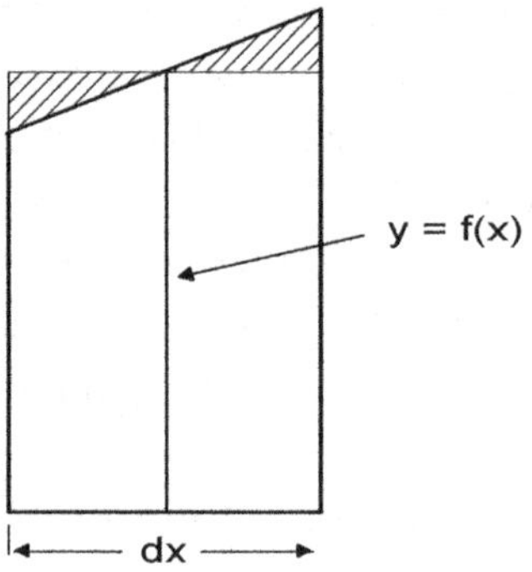

9-12 Area equals f(x) dx

The area of ... ima ... micro- region is thus equal ... the ... of th micro-areas must b ... as d ... very small, to the macro ... nder ... We indicate this in equ ...

area ... e =

The symbol, ∫, is know as the "integral" or "integration" sign. It indicates that we are summing up a vast number of infinitely small areas or elements. That is, we are essentially *integrating* all of the minute quantities together to form a macroscopic whole.

Calculus, including the operations of differentiation and integration, was introduced *ca. 1665* by Newton, and independently by Leibniz. However, it was not until *1854* (nearly *200* years later) that **Georg F. B. Riemann** (*1826-1866*) proved by means of the **Riemann sum** that the sum of the micro-rectangles was equal to the definite integral between *a* and *b* under the curve generated by the function.

This conclusion is given in form of the equation,

$$\lim_{n\to\infty} \sum_{k=1}^{n} f(x_k^*)\ \Delta x = \int_a^b f(x)dx$$

This equation indicates that as the number of "micro-rectangles" becomes infinitely large, the summation of their areas becomes equal to the integral function, regardless of where the "height", given by $f(x_k^*)$, may be located within the subinterval, Δx. In the function $f(x_k^*)$, the subscript, *k*, indicates that the function involves a particular *k* micro-rectangle of the infinite series of micro-rectangles. The superscript, * , denotes a single value of *y* (as calculated from $f(x_k^*)$) that occurs in the Δx subinterval in question.

Fundamental theorem of calculus

The equation of more interest in this tutorial, however, is the "fundamental theorem of calculus",

$$\int_a^b f(x)\,dx = F(b) - F(a)$$

In this equation, *F(b)* is the value of the integral *F* at $x = b$, and *F(a)* is the value of integral *F* at $x = a$. These values are just the *numbers* that we obtain when we evaluate the functions, *F(b)* and *F(a)*, and the *difference* between them gives us the value of $\int_a^b f(x)\,dx$. This is the process that allows us to calculate the value of integrals.

As an example, consider the function

$$f(x) = x^3 - 6x + 11x - 4$$

plotted in Fig. 9-13. We have then that

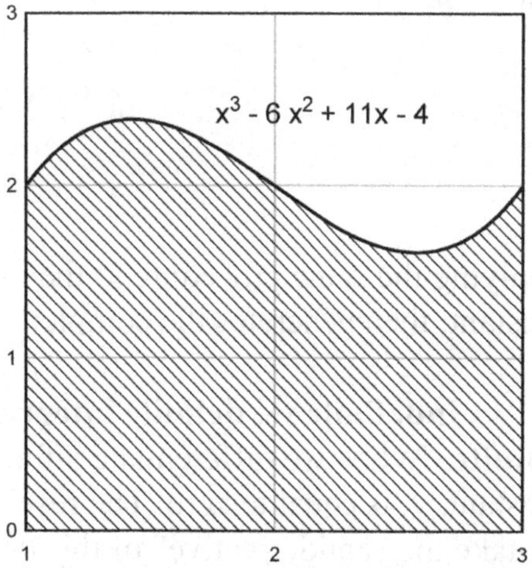

Fig. 9-13 Area under curve from x = 1 to x = 3

$$F'(x) = f(x),\ so\ F(x) = \int x^3 - 6x^2 + 11x - 4$$

$$= \frac{1}{4}x^4 - \frac{6}{3}x^3 + \frac{11}{2}x^2 - 4x + C$$

and then to evaluate the integral, we subtract *F(1)* from *F(3)*, during which the constants of integration, *C*, "subtract out", and we find

$$F(x{=}3) = \frac{1}{4}3^4 - \frac{6}{3}3^3 + \frac{11}{2}3^2 - 4\times 3 + C$$

$$= 20.25 - 54 + 49.5 - 12 = 3.75 + C$$

$$-\ F(x{=}1) = -(0.25 - 2 + 5.5 - 4 + C) = -\frac{(-0.25)}{4.00} - C$$

$$area\ under\ curve = 4\ units^2$$

And in Fig. 9-13, we see that the area under the curve is equal to *4 units*^{2}, which confirms that the integral, *F(x)*, yields the correct measure for this area.

The key in integration is that the function, *f(x)*, is the *derivative* of the integral, *F(x)*. So the integral, *F(x)*, must be such that, when differentiated, it gives the function, $x^3 - 6x^2 + 11x - 4$.

The function, *f(x)*, is just a series of power terms, and the rule for differentiating a power term is

$$if\ f(x) = x^n,\ then\ f'(x) = n\,x^{n-1},\quad and\ so\ d(x^4) = 4\,x^3$$

If we want the first term of the *derivative* of *F(x)* to be just x^3, then we must multiply the first term of *F(x)* by *1/4*, i.e.,

$$d(\frac{1}{4}x^4) = \frac{1}{4}d(x^4) = \frac{1}{4}\times 4\,x^3 = x^3$$

Applying the same procedure to the other terms of the function, and adding the integration constant, then generates the complete integral.

Indefinite vs. definite integrals. Integrals are spoken of as being either **indefinite** or **definite**. When we differentiate a function such as $x^3 - 6x^2 + 11x - 4$, the derivative of the constant, *4*, is equal to zero. The resultant derivative is thus $3x^2 - 12x + 11$. However, we can not take the "antiderivative" of this derivative to find the original function. Reversing the laws of differentiation lets us find the first three terms, x^3, $-6x^2$ and $11x$, but *no* operation on zero can tell us what the original constant was (or if it existed). We must write therefore that

$$\int (3x^2 - 12x + 11)dx = x^3 - 6x^2 + 11x + C \text{ (the integration constant)}$$

This integral is called the **indefinite integral** because it can not be "known" unless *C*, the integration constant, is known. Generally *C* is determined by finding the value of the integral for some certain set of circumstances (which is usually for the case where *x* equals *0*).

However, when we evaluate the integral by subtracting *F(a)* from *F(b)*, the *C* terms subtract out (as above), and the integral has a known value, and is thus a **definite integral**.

Further derivative-integral calculations. We can move further up-and-down the "derivative-integral" chain. If we treat the function above, *f(x)*, as a new *integral, F(x)*, then we can differentiate that integral to obtain a new function, $3x^2 - 12x + 11$; that is

$$\text{Let } F(x) = x^3 - 6x^2 + 11x - 4$$

$$\text{then } F'(x) = f(x) = 3x^2 - 12x + 11$$

A function generates a curve, and we have plotted the "new" *f(x)* curve in Fig. 9-14. This curve extends into the negative region of the graph.

The reason for this can be seen by comparing Fig. 9-14 with Fig. 9-13. The value plotted in Fig. 9-14 is the tangent *slope*, at each point, of the curve in Fig. 9-13. The slope is positive (cf. Fig. 9-13) for $x = 1$ to $x = 1.42$ but it is then negative up to $x = 2.58$ where it again becomes positive.

The precise points at which the positive-negative transition in slope occurs (where $y = 0$ in Fig.9-14) are the *roots* of the quadratic equation, $3x^2 - 12x + 11 = 0$, and can be calculated by **quadratic formula**. Also, the minimum in the curve in Fig. 9-14 at $x = 2$ corresponds to the point of *inflection* in the curve in Fig. 9-13, where the curvature changes from negative to positive.

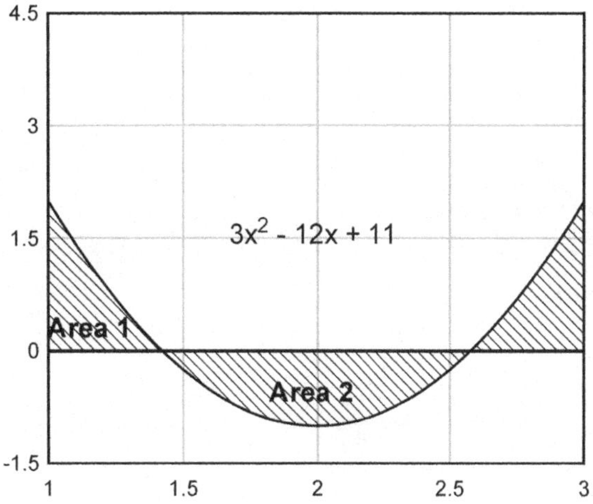

Fig. 9-14 Determining area under curve

However, we can *not* find the area "under the curve" in Fig. 9-14 simply by integrating *F(x)* from $x = 1$ to $x = 3$. This integration gives

$[F(x)$ [illegible] $6x$ [illegible]

$=$ [illegible] $\times(1)$ [illegible] $] = 6$

$=$ [illegible] $\times(1)$ [illegible] $] = \underline{-6}$

area [illegible] *ve* $= 0$

The area u[illegible] is cl[illegible] qual [illegible] discrepancy results from the fact that we can no[illegible] area [illegible] *ou*[illegible] 9-13, the region "under the curve" is bounded [illegible] and x [illegible] dic[illegible] ction curve at the top, and the x axis at the bottom. [illegible] n ther[illegible] cu[illegible] integral. In Fig. 9-14, the areas from $x = 1$ to [illegible] d from [illegible] x [illegible] ay; they lie above the x axis and are bounded. And [illegible] calc[illegible]

[illegible] $= |$ [illegible] -1 [illegible]

$=$ [illegible] $(1.4$ [illegible] $42) = 6.385$

$=$ [illegible] $(1)^2$ [illegible] $= \dfrac{-6}{0.385}$

[illegible] 0.38 [illegible] *nd* [illegible] $1 = 0.770$ *units*^{2}

The sum of the area [illegible] axis [illegible] al to [illegible] s^2.

However, th[illegible] y the [illegible] en [illegible] *below* the x axis is *not* bounded, and in fa[illegible] $y = -$[illegible] can [illegible] a 2 which lies *above* the curve, and is bounded abo[illegible] xis. In [illegible] betw[illegible] 42 and $x = 2.58$ gives

[illegible] $x^3 -$ [illegible] $\Big|_{1.42}^{2.58}$

[illegible] $2.58)^3$ [illegible] $)^2 +$ [illegible] 5.616

[illegible] $[(1.42$ [illegible] $42)^2$ [illegible] $)] = \dfrac{-6.385}{-0.769 \text{ units}^2}$

This straightforwar[illegible] indic[illegible] rea [illegible] *egative* 0.769 units^2.

This appare[illegible] n is t[illegible] wa[illegible] books. One is to convert all areas to their *absol*[illegible] This m[illegible] as [illegible] e other way is to consider areas below the x axis as [illegible] n tw[illegible] The ar[illegible] two curves is found by

$$\textit{Area between two curves} = \int_a^b f(x)\,dx - \int_a^b g(x)\,dx$$

One subtracts the area produced by the integration of the function, *g(x)*, from that produced by the integration of the first function, *f(x)*, to determine the area between the curves.

In the present case, the top "curve" is taken as being the *x* axis, whose integral is equal to *0* (it has no area). From this is subtracted the integral of the second function, i.e.,

$$\Delta\ Area = 0 - \left[x^3 - 6x^2 + 11x\right]_{1.42}^{2.58}$$

$$= 0 - (-0.769) = 0.769\ units^2$$

The area between the curves (Area 2) is then positive, and when added to the sum of the areas above the *x* axis (*2* × Area 1 or *0.770* units^2) gives a total enclosed area of *1.539* units^2. The concept that an area *always* lies between two curves (even if one is the *x* axis) is very useful in determining areas by integration.

Methods of integration

In theory, one can find the integral of "any" function by applying the rules of differentiation in reverse. In practice this turns out to be difficult for many functions. Reviews on integration on the Internet list as many as *10* different methods for finding the integral of a function. The problem is so difficult that a great deal of effort has been spent to develop computer programs which can integrate functions automatically. And there are seemingly simple functions such as,

$$e^{-x^2}$$

that can *not* be integrated in terms of elementary functions. (See the *Wolfram Integrator*, and its *FAQ* section and Internet links, for a quick exposition of the complexity of high-level integration.)

In elementary calculus, we can only integrate expressions that have been *differentiated* from the function that we wish to reproduce as an integral. Consider the power term, x^n,

$$f(x) = y = x^n,\ and\ \frac{dy}{dx} = nx^{n-1} \quad \textit{(differentiation rule)}$$

$$e.g.,\ y = x^3,\ and\ \frac{dy}{dx} = 3x^2$$

And integration is then just the reverse of this differentiation, e.g.,

$$\int dy = a\int x^n dx + C = a\frac{1}{(n+1)}x^{(n+1)} + C \qquad (\textit{integration rule})$$

$$\text{e.g.,} \int y = 3\int x^2 dx = 3\frac{1}{(2+1)}x^{(2+1)} \qquad \text{so } y = x^3 + (C=0)$$

One of the easiest ways for students to visualize the relationship between the function, the differential, and the integral is by means of a "table of standard forms". Here is an abbreviated example of such a table.

Table of Standard Forms

Differential, $f'(x) = \frac{dy}{dx}$	*Function*, $f(x) = y$	*Integral*, $F(x) = \int f(x)dx$
1	x	$\frac{1}{2}x^2 + C$
0	a	$ax + C$
1	$x \pm a$	$\frac{1}{2}x^2 \pm ax + C$
a	ax	$\frac{1}{2}ax^2 + C$
nx^{n-1}	x^n	$\frac{1}{n+1}x^{n+1} + C$
$-x^{-2}$	x^{-1}	$\ln x + C$
$\frac{du}{dx} \pm \frac{dv}{dx} \pm \frac{dw}{dx}$	$u \pm v \pm w$	$\int udx \pm \int vdx \pm \int wdx$
	Exponents and Natural Logs	
e^x	e^x	$e^x + C$
x^{-1}	$\ln x$	$x(\ln x - 1) + C$
$a^x \ln a$	a^x	$\frac{a^x}{\ln a} + C$
	Trigonometric Functions	
$\cos x$	$\sin x$	$-\cos x + C$
$-\sin x$	$\cos x$	$\sin x + C$
$\sec^2 x$	$\tan x$	$-\ln \cos x + C$

Most of the relationships in this Table are understandable to us because of the section above on differentiation. A few are unfamiliar, however, such as why does

$$\int \tan x \, dx = -\ln \cos x + C$$

We can explain this by using the **substitution rule** for integration. The substitution rule and the **integration by parts rule** are two commonly used techniques for integration. They are interesting for our purposes because they relate to the **chain rule** and **multiplication rule** which were introduced in the section on differentiation.

To evaluate the integral of *tan x*, we can use substitution and proceed as follows:

$$\textit{let} \int \tan x \, dx = \int \frac{\sin x}{\cos x} dx \quad (\textit{from trig identity of } \tan x)$$

Then we use the substitution $u = \cos x$, from which we have

$$u = \cos x$$

$$\textit{and differentiation gives} \quad \frac{du}{dx} = -\sin x$$

$$\textit{and then} \quad dx = -\frac{1}{\sin x} du$$

Inserting the identities from above for *cos x,* and for *dx,* into the original integral yields

$$\int \tan x \, dx = \int \frac{\sin x}{u}\left(-\frac{1}{\sin x} du\right) = -\int \frac{1}{|u|} du = -\ln|u| + C = -\mathbf{\ln|\cos x| + C}$$

In the differentiation section, we found that *dx/x* is the differential of *ln x*, so we know that the integral of *dx/x* is therefore equal to *ln x*. Moreover, since the logarithm of a *negative* number can not be determined, we specify the absolute value of *cos x* which is always positive.

Substitution is useful also with power terms or expressions. Consider the integral,

$$\int 6x(x^2 + 6)^3 \, dx$$

Let us make the substitution

$$u = x^2 + 6$$

$$\textit{then } du = 2x\,dx$$

Our original integral [illegible] dx. [illegible] we note that $6 = 3\times 2$, and if we move the constant, [illegible] of the [illegible] have

$$3\int(x^2 \ldots) = 3 \ldots = \frac{3}{4}(x^2 + 6)^4 + C$$

The first constant, [illegible] to ind[illegible] is [illegible] the final constant, C, but is actually equal to the [illegible] ant C [illegible] 3.

To verify th[illegible] can d[illegible] the [illegible] the chain rule (pg. 172). We define the "outer fu[illegible] eing

$$\ldots = \frac{3}{4} \ldots + C$$

[illegible]

The "inner function" i[illegible]

[illegible] $+ 6$

[illegible]

When these two di[illegible] "chai[illegible] obtai[illegible]

$$\frac{dy}{dx} \ldots (u) \ldots \quad \text{so} \quad \frac{dy}{dx} = 6x(u)^3$$

$$\textit{since} \ldots, \quad \frac{dy}{dx} \ldots 6)^3, \quad \ldots = \int 6x(x^2 + 6)^3\,dx$$

This integral is the [illegible] original [illegible] sult [illegible] And we see also that the method of *substitu*[illegible] ation i[illegible] the *chain rule* in differentiation.

The second [illegible] integr[illegible] rati[illegible], derives from the **multiplication rul**[illegible] iatio[illegible] i.e.

$$\ldots = u \ldots (uv) \ldots v\cdot du$$

where u and v are functions of x, and could be written $u(x)$ and $v(x)$. If we integrate both sides of the multiplication rule, and rearrange, we have

$$\int u\,dv = \int uv + \int v\,du$$

$$\textit{and since } \int dy = y, \textit{ if } y = uv, \textit{ then } \int d(uv) = uv$$

$$\textit{so } \int u\,dv = uv - \int v\,du$$

This is the "integration by parts" formula.

Here are some examples using the integration by parts technique. Let us solve for the integral

$$\int x \cos x \, dx$$

using the given formula

$$\int u\,dv = uv - \int v\,du$$

To do this, we must take the following steps,

$$\textit{set } u = x, \textit{ and } dv = \cos x \, dx$$

$$\textit{note that } v = \int \cos x \, dx = \sin x, \textit{ and } \int \sin x \, dx = -\cos x$$

$$\textit{then substitution into } \int u\,dv = uv - \int v\,du \textit{ gives}$$

$$\int x \cos x \, dx = x \sin x - \int \sin x \, dx \; (= -\cos x)$$

$$= x \sin x + \cos x + C \; (\textit{integral sought})$$

The complexities of determining derivatives and integrals may cause students to lapse into thinking that we are "only playing games with mathematical symbols." But in fact the expressions, although they may look strange, have "hard numbers" meanings.

In Fig. 9-15, we have this situation:

$$f(x) = x \cos x$$

$$f(x)' = \cos x - x \sin x$$
(derived separately)

$$\text{and } F(x) = \cos x + x \sin x$$
(from above)

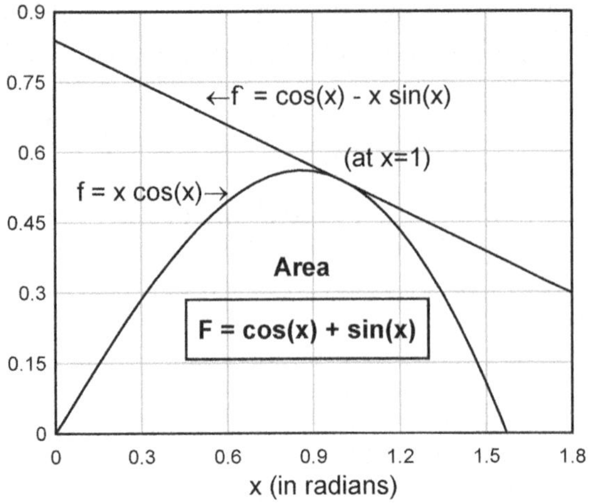

Fig. 9-15 Roles of f(x), f'(x) and F(x) for f(x) = x cos(x)

The function, *f(x)*, determines the course of the curve, as we normally see. The derivative of the function determines the instantaneous slope of the tangent line at a given value of *x*. For *x = 1 (in radians)*, the derivative defines the slope to be

$$f'(x) = \cos(1 \; radian) - (1)\sin(1 \; radian)$$

$$= 0.540 - (1)0.841 = -0.301 \; (slope \; of \; tangent \; line)$$

And if a line of this slope is translated upward along the *y* axis, it will be found to be just tangent to the curve at the point *x = 1*.

The area under the curve is given by

$$F(x) = \cos(x) + x \sin(x)\Big|_0^{1.57}$$

$$= \cos(1.57) + 1.57 \sin(1.57) = 0 + 1.57 \,(1) = 1.57$$

$$- (\cos(0) + 0 \sin(0)) = - 1 + 0 = \underline{- 1}$$

$$0.57 \; sq. \; units$$

The value of each rectangle on the graph is *0.3 x 0.15 or 0.045 sq. units*. Dividing the total area of *0.57 sq. units* by *0.045 sq. units* per graph rectangle indicates that the area under the curve should be equal to *12.67* graph rectangles.

It is difficult to count the number of rectangles under the curve with high accuracy, but an estimate of the area of the whole and partial rectangles under the curve suggests that it is equivalent to about *12.7-12.8* rectangles in total. The value of the integral is therefore essentially equal to the area under the curve.

The integration by parts technique is also useful with exponential functions such as

$$\int xe^x dx$$

Here we set the parts to be

$$u = x \qquad dv = e^x dx$$

$$and\ du = dx \qquad v = e^x$$

$$which\ we\ substitute\ into\ \int u dv = uv - \int v du + C$$

This gives the expression

$$\int xe^x dx = xe^x - \int e^x dx\ (equals\ e^x)$$

$$= xe^x - e^x = e^x(x - 1) + C$$

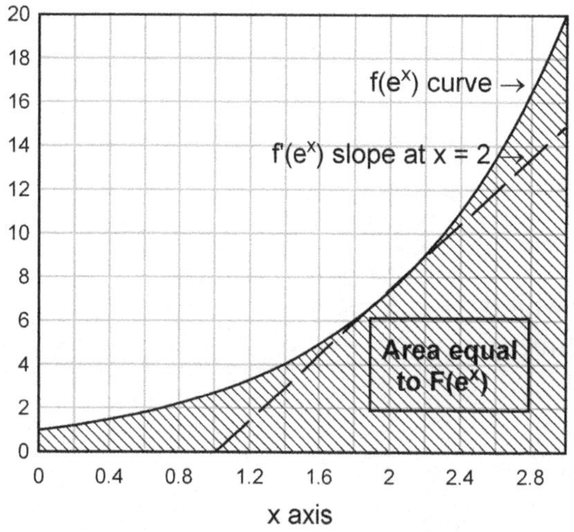

Fig. 9-16 Roles of f(x), f'(x) and F(x) for e^x

which we could solve easily because we know that $\int e^x dx$ is just equal to e^x itself (from the Table of Standard Forms, pg. 188).

The Table of Standard Forms shows that the function, derivative and integral for e^x *all* have the same value, just e^x. One might wonder if e^x alone can satisfy these three roles in calculus. But a test, as in Fig. 9-16, shows that in fact it can.

The curve in Fig. 9-16 is generated by the function, $f(e^x)$, just as usual in calculus. The slope of the tangent to the curve at, say, $x = 2$ is given by the value of e^2 which is *7.39*. And if we translate a dashed line of this slope upward along the $x = 2$ isopleth in Fig. 9-16, we find that indeed it forms a tangent to the curve, as is normal for the derivative, $f'(x)$. And we evaluate the integral also in the customary way, i.e.,

$$Area\ under\ curve = F(e^x) = e^x\Big|_{x=0}^{x=3} = e^3 - e^0 = 20.08 - 1 = 19.08\ sq.\ units$$

To check this answer, we count the number of squares in the cross-hatched areas, which is about *48* (including estimated partial squares). The area of each square is *2 x 0.2 = 0.4 sq.*

units. And the tota[illegible] is [illegible] This agrees well with the area of *19.08 sq. units* calc[illegible] the in[illegible] we see that the integral does give the area under [illegible]

In the prece[illegible]s, we [illegible] fferentiation and integration, and especially at th[illegible]s, as [illegible] grap[illegible] ccur between the various functions, their der[illegible]ntan[illegible]), [illegible]grals (area under the curve). And we have giver[illegible]ples o[illegible] qu[illegible] fferentiation and in integration. The Internet contai[illegible] suc[illegible], as [illegible] textbooks. Two web sites, *Calc101* and the *W*[illegible]*rator*, [illegible] n and integration by computer, may also be of inter[illegible]s lea[illegible]lus.

Multiple [illegible]

A function m[illegible]ore th[illegible] independ[illegible]e, i.e., $f(x,u,v...)$. Such a function is differen[illegible] **partial** dif[illegible]ion [illegible] where all of the variables are held constant, exce[illegible] the func[illegible]n d[illegible] with respect to that one variable. This prod[illegible]s of *pa*[illegible]ential [illegible] summed up to yield the *total* differential of the fu[illegible].

$$\text{if} \; [\text{illegible}] \; \text{then} \; [\text{illegible}] \; du + \left(\frac{dy}{dv}\right)_u dv$$

$$\text{or} \quad \text{total } d[\text{illegible}] \; \text{partial} [\text{illegible}] \; \text{differential}(2) + \cdots$$

The **multip**[illegible] also k[illegible] **it**[illegible] again") **integral**, is the way in which a function o[illegible] independ[illegible]iable[illegible]), can be integrated. Multiple integral functions a[illegible] the f[illegible]

$$[\text{illegible}] \, dx = [\text{illegible}] \, (x,y)\, dy \Big]_c^d dx$$

Let us integ[illegible]ble in[illegible], [illegible] its $x=1$ to $x=3$, and $y=2$ to $y=4$. The integral would [illegible]

$$[\text{illegible}] \text{ble int}[\text{illegible}] = \int_1^3 \int_2 [\text{illegible}] \, dx$$

This double integra[illegible] first [illegible] x [illegible] constant, and integrating the "..y^2dy" portion of [illegible] integra[illegible] using [illegible] of the *y* integral as a constant and integrating the [illegible], that i[illegible]

$$first,\ x^2\int_{y=2}^{y=4} y^2\,dy = x^2\left[\frac{1}{3}y^3\right]_2^4 = \frac{1}{3}(4^3 - 2^3)x^2 = \frac{56}{3}\ (= value\ of\ y)\,x^2$$

$$then,\ \frac{56}{3}\int_1^3 x^2\,dx = \frac{56}{3}\left[\frac{1}{3}x^3\right]_1^3 = \frac{56}{6}(3^3 - 1^3) = \frac{(56)(26)}{6}\ (double\ integral\ value)$$

Multiple integration is like partial differentiation in the sense that we treat all of the variables as a constant except one, and then "work", each in turn, with that one variable.

Multiple Integration over a Region

When we integrate $y = f(x)$, we obtain the area under the curve, with that area being bounded at its lower level by the x axis. If we integrate a function having two independent variables such as $z = f(x,y)$, we obtain a *volume* which is bounded at its lower limit by a **region (R)** which lies in the xy-plane.

Double integrals are often described as being calculated over a certain region in the xy-plane. Take the case where the region is a rectangle, whose area is given by $A = xy$. Since x and y are independent variables, the integral is

$$A = \int dA = \int_c^d\int_a^b f(x,y)\,dxdy$$

The rectangle is divided into an infinite number of micro-rectangles where the area of each micro-rectangle, dA, is equal to $dxdy$. Integrating the areas of all of the micro-rectangles *over* the whole region then yields the area of the macro-rectangle.

Notice that the general form for the double integral includes the function symbol, $f(x,y)$. For determining a surface area, however, $f(x,y)$ is simply equal to *one* because the integration of $dxdy$ gives $\int dx = x$, and $\int dy = y$, with then $xy = A$.

Let us determine the area of a rectangle of 3×2 units as a multiple integral, i.e.,

$$Area = \int_0^3\int_0^2 dydx = \int_0^3\left[\int dy = y\right]_0^2 dx = \int_0^3 (2 - 0)dx$$

$$or\ 2\int_0^3 dx = 2[x]_0^3 = 2(3 - 0) = 6\ sq.\ units$$

We first integrate y from *0 to 2* which yields the *constant, 2*. This constant is then placed in front of the integrand for x from *0 to 3*, which itself yields a constant of *3*. Multiplying 2×3 then gives the area of the rectangle under study.

If we can express y as a function of x, we can calculate the area of any two-dimensional figure as a double integral. Let the region be bounded between $y = 0$ and $y = x^2 + 3x$, and between $x = 0$ and $x = 3$, then we have

$$A = \int_0^3 \left[\int_0^{x^2+3x} dy \right] dx = \int_0^3 \left[y \right]_0^{x^2+3x} dx = \int_0^3 (x^2 + 3x) dx$$

$$= \left[\frac{1}{3}x^3 + \frac{3}{2}x^2 \right]_0^3 = 22.5 \; sq. \; units$$

In **three-dimensional plotting** (pg. 175), we commonly set z to be a function of x and y, or $z = f(x,y)$. In this case, for each micro-element of $dxdy$ on the xy-plane there is a *single* value for z which is calculated from the values of x and y *at* the micro-element in question. The product of $z \times dxdy$ gives the volume of the rectangular micro-prism that rests on the area $dxdy$. An integration of all of the micro-prisms that lie above the region in the xy-plane then yields the total volume of the body above this region. This is the *geometric* representation of the product of a double integration. It is similar to the product of a single integration, but with one more dimension added.

As an example of volume determination by double integration, let us consider the square pyramid. The volume of a square pyramid is known from geometry to be $V = 1/3hA$, where h is the height of the pyramid, and A is the area of its base. For a pyramid with $h = 2$, and each side of the square base, b, equal to $\sqrt{2}$, the volume is $1/3 \text{ x } 2 \text{ x } (\sqrt{2})^2 = 4/3$ cubic units.

We will find the volume of the pyramid by using a double integral to calculate the volume under *one* side of the square pyramid, and then multiplying by four. Let the pyramid be centered at $x = 0, y = 0$. The *plane* of the side lying lies in the first octant will then intersect the x,y,z axes at the points, $1,0,0$; $0,1,0$; and $0,0,2$, respectively. The equation for this plane is

$$z = 2 - 2x - 2y \; (\textit{from geometry})$$

This gives us z as a function of x and y, or $z = f(x,y)$. We now need to know y as a function of x, or $y = f(x)$. We can find this from the *line* that is generated when the pyramid plane intersects the x,y-plane; i.e., where z is equal to *zero*. This gives

$$\textit{if } z = 0, \textit{ then } 2 = 2y + 2x, \textit{ or } \quad y = 1 - x$$

We can then set up, and solve, the double integral,

$$V = \int_0^1 \int_1^{1-x} (2 - 2x - 2y)\,dydx$$

$$= \int_0^1 \left[2y - 2xy - 2\frac{1}{2}y^2 \right]_1^{1-x} dx$$

$$and, \ (2(1-x) - 2x(1-x) - \frac{2}{2}(1-x)^2) - (2(1) - 2x(1) - (1)^2)) = x^2$$

$$and \quad \int_0^1 x^2 dx = \left[\frac{1}{3}x^3 \right]_0^1 = \frac{1}{3} \ cubic \ units$$

Double integration indicates the volume of *one quarter* of the pyramid to be *1/3* cubic units, or *4* × *1/3* = *4/3* cubic units for the whole pyramid. This is same volume as given by the geometry formula.

Calculus textbooks and Internet sites give many examples of double integration. One popular exposition is to show how the volume of complex shapes can be calculated easily as the difference between the double integrals of two different volume functions, i.e.,

$$V_{complex} = V_g - V_f = \int_c^d \int_a^b g(x,y)\,dydx - \int_c^d \int_a^b f(x,y)\,dydx$$

This is an extension of the method for finding the area between two curves (pg. 187) by subtracting the integrals of the two functions, i.e.

$$Area \ between \ two \ curves = \int_a^b f(x)\,dx - \int_a^b g(x)\,dx$$

Double integrals are thus just one integral being evaluated after another. Each individual integration is no harder (nor for that matter easier) than the integration of a single-variable function. The carrying-forward of the value of the first variable as a function of one or more following variables can lead to tedious calculations, however, especially for triple and higher integrals.

Epilogue

In this book [illegible] to em[illegible] "See [illegible] method of teaching children to read. "See Spot ru[illegible] poken [illegible] es minimum information, and of itself, teaches noth[illegible] ar, o[illegible] structure. These will come later. But it gives [illegible] that [illegible] use confidently. And, I would argue, all of literac[illegible] ved fr[illegible] extensions of "See Spot run." Besides, kids do lik[illegible] t run.

It is my hop[illegible] hat [illegible] g of the same aid to beginning math students. Ma[illegible] not th[illegible] tual ability; some students are good at math, whil[illegible] ss in[illegible] tain level of knowledge about mathematics is usef[illegible] oft[illegible] ly not that hard — after some "boning up" on the [illegible] con[illegible] math[illegible] quations — to obtain.

Made in the USA
Monee, IL
05 November 2026

67375699R00118